Luis Rivacoba

New plant nitrogen measurements for fertilizer recommendations

Luis Rivacoba

New plant nitrogen measurements for fertilizer recommendations

Adaptation to climate change

ScienciaScripts

Imprint

Any brand names and product names mentioned in this book are subject to trademark, brand or patent protection and are trademarks or registered trademarks of their respective holders. The use of brand names, product names, common names, trade names, product descriptions etc. even without a particular marking in this work is in no way to be construed to mean that such names may be regarded as unrestricted in respect of trademark and brand protection legislation and could thus be used by anyone.

Cover image: www.ingimage.com

This book is a translation from the original published under ISBN 978-620-0-01608-9.

Publisher:
Sciencia Scripts
is a trademark of
Dodo Books Indian Ocean Ltd. and OmniScriptum S.R.L publishing group

120 High Road, East Finchley, London, N2 9ED, United Kingdom
Str. Armeneasca 28/1, office 1, Chisinau MD-2012, Republic of Moldova, Europe
Printed at: see last page
ISBN: 978-620-6-97745-2

ACKNOWLEDGMENTS

I would like to express my gratitude to all those who have contributed to the development of this Doctoral Thesis and who in one way or another have made its elaboration possible.

First of all, I would like to thank Dr. Alfonso Pardo Iglesias for his rigorous and enriching scientific assessments, which have served me as a guide, encouragement and orientation during this process.

To the colleagues who have been part of the Natural Resources section of SIDTA-CIDA, Dr. Mª Luisa Suso, Dr. Nuria Vazquez and Dr. Leticia Olasolo, for their timely advice and corrections. To Pilar Yecora and Cristina Casis for their commendable work in the laboratory, and to all the staff of the center who have collaborated in the work and facilitated the daily tasks, especially the operators. I would also like to thank Ma Carmen Arroyo and her team at the Regional Laboratory for their thoroughness and speed in analyzing the soil samples.

To the Servicio de Investigacion y Desarrollo Tecnologico Agroalimentario de La Rioja (SIDTA), where this doctoral thesis was carried out, for facilitating its development. To Dr. Enrique Gartia-Escudero and all the staff of SIDTA, who directly or indirectly participated in this work. To the Consejena de Agricultura, Ganadena y Medio Ambiente del Gobierno de La Rioja, for allowing the use of the facilities and equipment of SIDTA-CIDA.

To the Instituto Nacional de Investigation y Tecnologia Agraria y Alimentaria (INIA) for granting me the predoctoral scholarship linked to the national project "Integration of soil, plant and simulation models for the efficient management of nitrogen in horticultural crops" (RTA-2011-00136-C04-02) that has made it possible for me to carry this out.

And of course, finally, I would like to thank my parents, who are always there, and my sister, for their sacrifices, patience and valuable teachings, because without them this would not have been possible. To my grandfather, who is no longer here but who was always there, for teaching me that struggle, effort and sacrifice, with a smile on your face, are the way to achieve your goals. To Sara, for watching me and understanding me, for the infinite support. To all those who directly or indirectly have facilitated the completion of this work, thank you very much.

<u>SUMMARY</u>

There is a growing interest in optimizing nitrogen fertilization of crops and improving nitrogen use efficiency to obtain high yields and limit side effects related to nitrogen leaching. Nitrogen concentration in plants is usually determined by chemical analysis, although there are alternatives such as $N\text{-}NO_3$ determination⁻ in sap and chlorophyll measurement as an indirect estimate of nitrogen content. Currently, reflectance, transmittance and fluorescence methods in specific regions of the spectrum are used to diagnose the nitrogen nutritional status of plants. The first objective of this work was to study the effect of available nitrogen on yield and nitrogen use efficiency in a cauliflower crop (*Brassica oleracea* var. *botrytis*) of Barcelona, Typical and Casper varieties. The second objective of this study was to evaluate the measurement of nitrate concentration in sap and the Nmin method, and the third objective was to evaluate the use of sensors based on the principles of reflectance, fluorescence and transmittance in the same test plots. The trials were carried out at Finca Valdegon, at the Centro de Investigacion y Desarrollo Tecnologico Agroalimentario del Gobierno de La Rioja, in Agoncillo (La Rioja), and at La Finca Experimental del Instituto Navarro de Tecnolog^a e Infraestructuras Agroalimentarias in Sartaguda (Navarra), Spain, in the years 2012, 2013 and 2014. The plantations were carried out in August. An experiment was designed with four treatments and four replicates in a randomized scenario depending on the mineral nitrogen content in the soil at the beginning of the crop. Ion exchange resins were installed in tubes at 0.2 m depth to estimate the mineralization of soil organic matter. Soil nitrogen, plant nitrogen, sap nitrate, resin nitrate, height, cover, biomass and leaf measurements were periodically sampled with SPAD®, Dualex®, Multiplex®, and Crop-Circle® sensors.

In the Barcelona variety, the mean value of available N above which no yield response was found was 184 ± 20 kgN/ha. In the Typical variety this amount was 189 ± 45 kgN/ha. In Casper the value found was 143 ± 7 kgN/ha, this value was significantly lower than in the rest of the trials, possibly due to an underestimation of the mineralization of soil organic matter.

The results of the balance analysis confirm the usefulness of the Nmin method for the design of nitrogen fertilization of cauliflower, as well as the importance that mineralized nitrogen, volatilization losses and correct irrigation scheduling can acquire in the balance to reduce leaching losses. The average extractions of the cauliflower varieties studied were 246 kg of nitrogen per hectare.

The mineralization of soil organic matter measured in the field reached an average value of 46 kg N/ha for the crop ratoon in the topsoil. The nitrogen supply through this process can account for up to 20% of the nitrogen extractions of the plant.

The concentration of N-NO3⁻ in sap has been a very sensitive and repeatable indicator, capable of showing significant differences between treatments, especially after top dressing fertilization.

For the NBI indices of the Dualex and Multiplex equipment and the REDVI index of Crop Circle, high correlations have been found with the nitrogen content of the plant, which has been used to adjust, depending on the phenological stage of the crop, functions that relate the measurements made with the biomass of the crop in the case of the NBI index and with the nitrogen nutrition index (NNI) in the case of the REDVI index. In the analytical curves obtained for the sensors, it can be observed that from a biomass of approximately 1 Mg/ha (50% of soil covered) the measurement stabilizes and it is from this moment on when deficit treatments begin to be distinguished from non-deficit treatments, as occurs with the total nitrogen content in the leaf. Therefore, these models can be used to determine a nitrogen nutritional deficit and to correct it by fertilization. It is very important that these deficiencies are detected in time so that they can be corrected.

The only method capable of quantitatively specifying a fertilizer recommendation today is the Nmin analysis. Delaying the soil analysis until moments before the top dressing fertilization, allows us to adjust more effectively the nitrogen fertilization, since we take into account the possible mineralization of the soil organic matter from the beginning of the crop, as well as the possible losses by leaching.

The results observed with the nondestructive measurement equipment complement the Nmin results and demonstrate the usefulness of these methods for detecting deficiencies in the nutritional status of plants. Of these, the CROP CIRCLE ACS-430 reflectance meter has allowed continuous analysis of a large number of samples in a short period of time, thus reducing the sampling error and obtaining more representative values of the crop than with the other equipment used.

The objective of these teams should focus not only on the early detection of nutritional deficiencies, but also on the quantification of these deficiencies in order to correct them quantitatively.

INTRODUCTION

The fruit and vegetable sector plays a very important role both in the agricultural sector and in the Spanish economy as a whole. In the advance published for the year 2015 (MAGRAMA, 2015), its participation in the Production of the Agricultural Branch reached 14,544 million euros, 34.1%, a highly significant figure that has followed a growing trend over the last few years. The area dedicated to horticultural crops in 2014 was 389,472 hectares with a total production of 14 million tons (MAGRAMA, 2015).

In La Rioja, during the 2014 campaign, the area dedicated to this type of crop was 4,625 ha, with the most representative crops being green peas, green beans, cauliflower and artichoke (Government of La Rioja, 2015).

In Spain, the consumption of inorganic fertilizers in the 2013/2014 campaign was 5.0 million tons, with 53.1% corresponding to straight nitrogen fertilizers (ANFFE, 2013/14). Fertilizers and amendments destined directly to agriculture, with a value of 2,030.4 million euros, represented in 2014 9.6% of the value of all intermediate consumption, and 5.0% in relation to the production value of the agricultural branch (MAGRAMA, 2015). In economic studies carried out by the Technical Institute of Agricultural Management of Navarra (ITGA) in different crops, it is observed that in most horticultural crops the cost of fertilizers represents between 5 and 10% of total crop costs, a low level compared to a crop such as corn for grain where they represent about 20%. This may cause farmers to over-fertilize to minimize the risk of a drop in productivity due to a lack of fertilizer.

Thus, the proper management of N in horticultural crops is of economic importance to the farmer but also affects the environment through its effect on groundwater pollution (EU Directive 91/676 on water pollution by nitrates of agricultural origin), and through the emission of nitrous oxide into the atmosphere by denitrification of soil nitrate.

In horticultural crops, nitrogen fertilization is usually high and this, together with the fact that they are irrigated crops and that in many cases the root system is relatively shallow, leads to high nitrate losses by leaching. The European Directive and the Action Programs derived from it, establish that the member states must dictate the maximum amounts of nitrogen fertilizer to be used in the different crops and the Codes of Good Agricultural Practices must be based on experimental studies in the different agricultural areas.

Regarding the impact of fertilization practices on greenhouse gas emissions, nitrogen fertilizers have been shown to be the main cause of the increase in N_2O emissions attributed to agricultural activity (Stehfest and Bouwman, 2006). Nitrous oxide is a gas produced in the soil as a result of denitrification or nitrification (Bremner, 1997) and has a global warming potential almost three hundred times greater than CO_2. It is estimated that 42% of total N_2O emissions to the atmosphere come from agriculture (IPCC, 2007). There is evidence that the best nitrogen fertilization practices are those that produce the highest agronomic efficiency, i.e. those that result in a higher rate of uptake of the applied N by the crop (Van Groenigen *et al.*, 2010). Thus, it is necessary to develop methods to determine, for each plot and each crop, the fertilization required to obtain high yields of good quality and, at the same time, with a reduced impact on the environment.

Nitrogen fertilization studies of horticultural crops are essential in order to be able to better adjust the optimum fertilizer doses. This adjustment is considered convenient due to economic and environmental reasons, and also because there is a great variation of these doses in the Action Programs of the different Autonomous Communities, a variation that is not easily justifiable (Ramos and Ubeda, 2009).

1.1. Cauliflower cultivation

Cauliflower (*Brassica oleracea* L. var. *botrytis* L.) belongs to the cruciferous family, which constitutes an important family of about three hundred genera, mainly herbaceous plants of temperate zones. The *Brassica* genus is the most economically important. Within the *Brassica* species, the species *B. oleracea* contains a number of plants that have become very important horticultural crops in temperate regions, such as smooth-leaved cabbage (*B. oleracea* var. *capitata* L.), Brussels sprouts (*B. oleracea* var. *gemmifera* DC.) or broccoli (*B. oleracea* var. *italica* Plenck).

In 2013 the production of cauliflower in Spain reached 146,700 t. and in 2014 in La Rioja, 9,518 t. were produced, making it one of the main vegetables produced in the Community where it also has a Protected Geographical Indication.

1.1.1. Periods of cauliflower development and crop cycles

The vegetative development of the cauliflower plant has been extensively studied, differentiating three vegetative phases during the commercial crop cycle (Wurr *et al.*, 1981). The first of these is the juvenile phase from germination to flower induction. In this phase only leaves are formed, varying in number between 12 and 20, depending on cultivar and temperature of the penode (Wiebe, 1975; Wurr *et al.*, 1981; Booij and Struik, 1990). The second corresponds to the floral induction phase, which is not clearly linked to the end of the juvenile phase (Booij and Struik, 1990). Induction is mainly caused by low temperatures and is also influenced by the age of the plants, variety, etc. The value of the vernalizing temperature varies according to the cultivar, from 6° to 8°C for winter cultivars to temperatures above 15°C for summer cultivars. The duration of vernalizing temperatures also varies with the cultivars, ranging from two to five weeks for autumn cultivars and from five to fifteen weeks for winter cultivars. At the end of this phase, leaf formation ceases. As there is a correlation between leaf number and pellet production, it is important to adjust the cycles so that induction occurs when the plant has a sufficient number of leaves. The third phase is the inflorescence growth phase (Wiebe, 1975; Wurr *et al.*, 1981).

According to this development pattern, the most commonly used cultivars in our country can be framed in the following productive cycles:

- Short cycle and harvesting with a duration between 45 and 90 days from planting to harvesting.

- Medium cycle and late autumn to mid-winter harvesting with a duration between 90 and 120 days from planting.

- Long cycle and harvesting from mid-winter to early spring. spring and a cycle duration between 120 and 180 days.

1.1.2. Cauliflower nitrogen fertilization

Cauliflower is a crop that is harvested when it is in full growth. Data on the dynamics of N uptake in this crop show an increasing line and suggest a continuous extraction until harvest (Everaarts, 1993).

Nitrogen withdrawals for this crop, according to several studies, can vary between 150 and 300 kg N/ha (Everaarts *et al.*, 1996), 170 and 250 kg N/ha (Everaarts, 2000) and 250 and 498 kg N/ha (Vazquez *et al.*, 2010).

According to Everaarts *et al.* (1996), the optimum nitrogen fertilization for this crop is around 224 kg N/ha minus the initial mineral nitrogen present in the soil. Csizinszky (1996) obtained the maximum production in green cauliflower with 294 kg N/ha of fertilizer. Rahn *et al.* (1998) determined the production ceiling for values between 240 and 300 kg N/ha of fertilizer. Rather *et al.* (2000) determined the optimum dose at 250 kg N/ha, as the sum of mineral N in the soil (Nmin) at transplanting and the N applied as fertilizer; these authors concluded that when the Nmin was higher than 210 kg N/ha in the soil horizon between 0 and 30 cm or higher than 270 kg N/ha in the soil profile from 0 to 90 cm, no response to nitrogen fertilization was found.

In trials conducted by Riley and Vagen (2003), no effect on yield was found in relation to increasing nitrogen fertilization from 150 to 250 kg N/ha. They also showed that split application of nitrogen gave significantly higher yields compared to a single application. Van Den Boogaard and Thorup-Kristensen (1997) found no response above 250 kg N/ha available (Nmin + fertilizer). A fertilizer increase of 100 kg N/ha led to an increase of 17 kg/ha of residual N in soil, 52 kg/ha of nitrogen in crop residues, 37 kg/ha of mineralizable N and 15 kg/ha of N in inflorescences. Nitrogen use efficiency increased when more nitrogen was supplied at the time of highest demand and growth.

In the recommendation of nitrogen fertilization in cauliflower using the Nmin method, Feller and Fink (2002) indicate as target Nmin 297 kg N/ha up to 0.6 m depth, considering 251 kg N/ha extracted by the crop and 40 kg N/ha as residue.

Nitrogen is not only a yield factor but is also related to quality. Thus, Rather *et al.* (1999) observed a significantly higher percentage of loose cauliflowers in nitrogen deficient states in relation to optimum fertilization. Bohmer *et al.* (1981) also observed an increase in the number of loose cauliflowers as a function of available nitrogen. An increase in nitrogen fertilization from 80 to 120 kg N/ha led to a 7% decrease in vitamin C content in cauliflower (Lisiewska and Kmiecik, 1996).

N losses from cauliflower residues can be high. Brassicas used in horticulture have a low harvest index (Abuzeid and Wilcockson, 1989; Everaarts and de Moel, 1991) and leave a large amount of nitrogen in the residues (Alt and Wiemann, 1990). In cauliflower, the harvest rate ranges from 33 to 47% according to several authors. Lorenz *et al.* (1989) reported that cauliflower residues contributed 130 kg N/ha to the following crop. Rahn *et al.* (1992) measured up to 300 kg N/ha in *brassica* residues. Everaarts (1993) determined values between 100 and 200 kg N/ha in cauliflower and Everaarts (2000) found values between 95 and 140 kg N/ha in cauliflower residues. Riley and Vagen (2003) determined that the residues left by cauliflower after harvest amounted to approximately 70 percent of the total. This organic nitrogen, after being mineralized, may be available for the next crop or may be leached, with consequent environmental risks. Scharpf (1991) found that 70% of the N in vegetable residues was available within ten weeks.

1.2. Nitrogen fertilization recommendation systems

Most nitrogen fertilizer recommendation systems for horticultural crops can be grouped into two basic categories. The first comprises those based on mineral N analysis at the beginning of the crop before plant growth is most rapid (Hartz *et al.*, 2000; Feller and Fink, 2002; Heckman, 2002) and the second includes those based on measurements on plants that indicate their nutrient status at certain times of their development (Hochmuth, 1994 and 2009; Rodrigo, 2006; Westerveld *et al.*, 2004; Rodrigo and Ramos, 2007a). Schroder *et al.* (2000) in a review of the two options conclude that the ideal diagnostic tool "should be able to detect both a deficiency and an excess of nitrogen, should be easy to handle and able to inform the farmer quickly about the need for supplemental nitrogen. The results obtained should be specific to the N status of the crop and the influence of other factors, i.e. variety, climatic differences between years, etc., should not affect the result".

In the last 20 years there has also been a growing interest in the use of simulation models for nitrogen fertilization decision support (Van der Burgt *et al.*, 2006; Rahn *et al.*, 2010a; Shaffer *et al.*, 2010).

1.2.1. Systems based on soil analysis

There are several nitrogen fertilizer recommendation systems based on soil measurements (Neeteson, 1995; Tremblay *et al.*, 2001; Hartz, 2002a). Among them, two types stand out: 1) those that determine the amount of mineral N that should be available for the crop at the beginning of cultivation at a certain soil depth, for example, the so-called Nmin system (Feller and Fink, 2002), and 2) the methods that determine the top dressing based on the soil analysis of the first 30 cm and that, depending on the values obtained, diagnose whether or not fertilization is likely to increase yields (Krusekopf *et al.*, 2002).

The Nmin (mineral nitrogen) method of nitrogen fertilizer recommendation was developed in Germany (Wehrmann and Scharpf, 1986). It is based on the determination of the mineral N content in the soil at the beginning of crop growth. In this method the optimum fertilizer dose is the difference between the total nitrogen required by the plant and the amount of mineral N found in the part of the soil explored by the root system at the beginning of the crop, and has been determined in different horticultural crops (Scharpf and Weier, 1996; Wehrmann and Scharpf, 1986).

Among the latter is the PSNT (Pre-sidedress Soil Nitrate Testing) system. This is a very useful tool to identify those plots where additional N application will increase yield. Although originally developed for corn (Magdoff, 1991), it has subsequently been adapted for broccoli, cauliflower, cabbage, celery, lettuce, sweet corn and tomato (Heckman *et al.*, 1995; Sanchez, 1999; Mitchell *et al.*, 2000; Hartz *et al.*, 2000; Breschini and Hartz, 2002; Krusekopf *et al.*, 2002). Top dressing takes place at least one month after soil preparation, during the previous phase of rapid plant growth when N demand is highest. Soil nitrate is considered to provide a rough idea of the availability of mineral N for the crop during the rest of the cycle (Hartz, 2002b). In this method, only the soil layer from 0 to 30 cm is usually sampled, however some authors recommend sampling the layer from 5 to 30 cm since the nitrate present in the first five centimeters is not very accessible to the roots. The nitrate content of the soil is used as an indicator of the need or not to apply top dressing. This system implies the need to be able to analyze soil nitrate quickly (Ramos, 2005).

All these systems need to be adjusted to different production regions with different climates and soils, and also with irrigation practices with varying efficiencies, as all these factors influence commercial production and also in different terms of soil nitrogen balance (Hartz, 2003). These systems are very useful in the case of horticultural crops because it is very common that the residual Nmin values in the soil from the previous crop are very high (Ramos *et al.*, 2002; Vazquez *et al.*, 2006) and, therefore, fertilization can be greatly reduced or even suppressed.

The main drawback of these systems is the cost of soil sampling and analysis and the high spatial variability of nitrate in the soil (Lopez-Granados *et al.*, 2002; Giebel *et al.*, 2006). In order to reduce the cost of the analysis, the following have been developed developed simple and economical methods (Hartz, 1994; Sepulveda *et al.*, 2003; Thompson *et al.*, 2009).

1.2.2. Systems based on plant measurements

To study the excess or deficit of nitrogen in the plant, the cntic nitrogen concentration can be used, which has been defined as the minimum concentration in the plant necessary to reach a maximum growth rate (Ulrich, 1952). Lemaire and Salette (1984) developed the concept of cntic nitrogen concentration in aerial biomass, at a given time of vegetative growth, as the minimum nitrogen concentration necessary to reach maximum biomass. They represented this concentration by a potential equation where total biomass is a function of nitrogen concentration, expressed as a percentage of total dry matter. Below the curve, growth is limited by nitrogen, above it is not limited, and above the curve, nitrogen concentration is optimal. It has been necessary to define the specific values of the coefficients of the nitrogen curve for the different species. For example, this has been done for forages (Lemaire and Salette, 1984), potatoes (Greenwood *et al.*, 1990), wheat (Justes *et al.*, 1994), corn (Plenet, 1995), green beans (Olasolo, 2013).

These dilution curves can be used to determine nitrogen requirements and to calculate the nitrogen nutrition index (NNI) which quantifies the nitrogen status of plants (Lemaire *et al.*, 1989; Lemaire and Meynard, 1997) and can be used in dynamic models to account for the effect of nitrogen on growth and yield (Justes *et al.*, 1997).

The general model of Greenwood *et al.* (1986) and a specific model for brassicas (Greenwood *et al.*, 1996) have been used to study the cnt curve in cauliflower. Riley and Vagen (2003) developed a cntic nitrogen curve model for broccoli and cauliflower that was closer to the general Greenwood equation than to the brassica-specific one. The nitrogen content of cauliflower was studied by Rincon *et al.* (2001) in Spain, obtaining a value of 6,959 kg/ha of dry biomass and an average nitrogen percentage of 4.6%, higher than the 4% predicted by the specific model for brassicas.

In general, measurements of plant nutrient status in relation to nitrogen are less laborious than soil measurements. The most common practice has been either foliar N analysis or sap nitrate analysis. However, there is a disparity of opinion among experts as some have found these measures useful (Kubota *et al.*, 1996; Rodrigo *et al.*, 2005; Hochmuth, 2009), while others have found soil analysis to be more effective than sap nitrate analysis in detecting when plants will respond to nitrogen fertilization (Pritchard *et al.*, 1995; Sanchez, 1998; Hartz, 2003).

Several studies indicate a close relationship between leaf chlorophyll concentration and leaf nitrogen content, because most of the leaf nitrogen is found in chlorophyll molecules (Evans, 1989; Peterson *et al.*, 1993).

Chlorophyll concentration, or leaf greenness, is affected by numerous factors, one of which is the nitrogen nutritional status of the plant. Therefore, by employing tools capable of measuring leaf chlorophyll content, we can improve the management of nitrogen fertilization in crops (Peterson *et al.*, 1993; Smeal and Zhang, 1994; Balasubramanian *et al.*, 2000).

The measurement of chlorophyll as an indicator of plant nutrient status is based on the good correlation observed in many crops between leaf chlorophyll content and leaf N content, especially when nitrogen deficiency is present (Schepers *et al.*, 1992; Samborski *et al.*, 2009). The sensors used in this measurement are easy to operate and relatively inexpensive (Goffart *et al.*, 2008). Some horticultural crops in which this type of measurement has been studied are: artichoke and romanesco (Rodrigo and Ramos, 2007c), tomato (Gianquinto *et al.*, 2006) and bell pepper (Godoy *et al.*, 2003). Chlorophyll measurement for nitrogen fertilizer management of horticultural crops has been reviewed by Rodrigo and Ramos (2007a), Gianquinto *et al.* (2004) and Tremblay (2013).

A fairly general problem with systems based on plant measurements is that these measurements are sensitive to other factors such as water deficits, nutrient availability (other than N), diseases, etc. In recent years, optical sensors have been developed to determine the N nutrient status of crops.

- **Sap analysis**

Rapid nitrate sap analysis is a practical and simple method for monitoring nitrate content in horticultural crops (Rodrigo and Ramos, 2007a). This method has been used to diagnose the nutritional status of the plant using specific sufficiency ranges for each crop (Kubota *et al.*, 1997). These measurements are affected by soil mineral N, plant petiole position, plant age, cultivar and time of day, but by selecting newly expanded mature leaves and sampling before noon, the variability in sap measurements can be reduced (Hochmuth, 1994 and 2009). Sap nitrate analysis has been used in several horticultural crops: cabbage (Scaife and Stevens, 1983), tomato (Prasad and Spiers, 1985; Hochmuth 1994 and 2009; Beverly, 1994; Thompson *et al.*, 2009), cauliflower (Kubota *et al.*, 1996), broccoli (Kubota *et al.*, 1997; Gardner and Roth, 1989), lettuce (Huett and White, 1992; Gartia *et al.*, 2003), bell pepper (Hartz *et al.*, 1993; Hochmuth, 2003), pumpkin (Studstill *et al.*, 2003), potato (Errebhi *et al.*, 1998; MacKerron *et al.*, 1995), artichoke (Rodrigo and Ramos, 2007b), carrot, cabbage and onion (Westerveld *et al.*, 2003) and others. This method is accurate, safe, convenient, simple and inexpensive (Prasad and Spiers, 1984). It has been described as a valuable and rapid technique for estimating nitrogen requirements (Kubota *et al.*, 1997) and has also been criticized for the high variability of its results and the lack of agreement of critical levels (Westerveld *et al.*, 2003).

- **Measurement of photosynthetic compounds in leaves by optical sensors**

At present, there are different optical sensors capable of measuring different photosynthetic compounds in the plant canopy, by reflectance, transmittance or fluorescence methods, capable of estimating the nitrogen nutritional status of plants.

The reflectance and transmittance methods are based on the following concept: a part of the radiation incident on the leaves is reflected specularly while the other part penetrates the leaf and is subjected to multiple scattering due to discontinuities in the refractive index between the cell walls and the air, and between the cell walls and the water inside the leaf tissue. A portion of the scattered radiation can escape through the lower epidermis of the leaves, receiving the name of transmitted radiation. The remaining radiation continues to undergo scattering processes within the leaf or escapes through the upper epidermis, which is called diffuse reflected light (Woolley, 1971), forming part of the total reflected radiation (Willstatter and Stoll, 1918; Wendlandt, 1966; Kumar *et al.*, 2001). According to the law of energy conservation, the sum of the fractions of absorbed, reflected and transmitted light must equal one (Lee and Graham, 1986).

The nature and amount of light reflected, absorbed or transmitted depend on the wavelength of the incident radiation and its angle of incidence, on the roughness of the leaf surface (Kumar *et al.*, 2001) and on differences in the refractive indices of the cuticle in the case of leaves with waxy cuticles (Hoque and Remus, 1996). In addition, they are influenced by the internal structure of the leaf, by the pigment content and its distribution within the leaf, and by the quantity and quality of chloroplasts. The angle of leaf exposure controls the diffusion or scattering and the optical passage of incident light (Hoque and Remus, 1996). Finally, the water content of the leaf, both the concentration and its distribution, controls the refraction index in the visible range of the electromagnetic spectrum and the absorbance in the near infra-red (Hoque and Remus, 1996).

Research in this area continued until the mid-1960s, when fundamental work relating the optical properties of plants to their morphological characteristics emerged. The work of Allen, Gates, Gausman and Woolley pioneered the use of visible and near-infrared radiation to obtain information on the reflectance, transmittance and absorbance of plants in different plant crops (Gates *et al.*, 1965; Allen and Richardson, 1968; Allen *et al.*, 1969; Gausman *et al.*, 1969; Gausman and Allen, 1973; Gausman, 1974; Woolley, 1971).

In the green zone (around 550 nm, Figure 1) and in the far red zone (around 700 nm, Figure 1) the reflectance is sensitive to chlorophyll variation. This is because the absorption of chlorophyll in the far red region of the electromagnetic spectrum (Figure 2) is high enough to allow light to penetrate deep into the leaf, unlike in the green region, (Gausman and Allen, 1973; Gitelson and Merzlyak, 1994) and therefore the reflectance at these wavelengths is maximum in the green region and minimum in the red region, allowing a very accurate assessment of chlorophyll content.

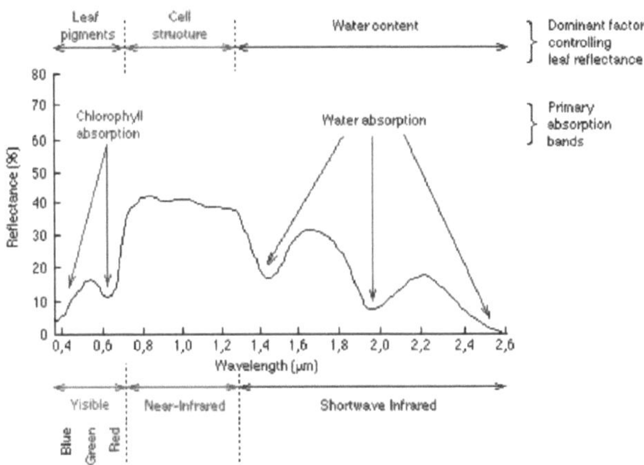

Figure 1. Electromagnetic spectrum. Wavelength expressed in pm. Source: Hoffer (1978).

The optical properties of leaves are the same regardless of the species involved. A healthy leaf has a spectral characteristic that differs in each of the three main regions of the spectrum. Figure 1 shows that in the visible band (400-700 nm) light absorption by leaf pigments dominates the generally low (at most 15%) reflectance spectrum of the leaf. There are two main absorption bands, in the blue (450 nm) and in the red (670 nm) due to the absorption of the two main pigments: chlorophyll a and b which constitute 65% of the total leaf pigments in higher plants.

In the near infrared region (700-1300 nm), the structure of the leaf explains the optical properties. Pigments and cellulose are transparent at these wavelengths and therefore the absorption of radiation is very small (of the order of 10% at most) but not the reflectance and transmittance which can reach up to 50%. The third important region corresponds to the infrared between 1300 and 2500 nm, characterized by the absorption of radiation by the water contained in the leaf.

In recent years, remote sensing research has relied on the use of spectral radiation, captured by field spectrometers and satellite and airborne radiometers (imaging spectroscopy), to determine the characteristics of materials and/or coverages on the earth's surface; these characteristics can be either the identification of a type of material (types of minerals, plant species), or the determination of a variable related to a type of cover (plant stress, phenological state): the identification of a type of material (types of minerals, plant species), or the determination of a variable related to a type of cover (plant stress, phenological state). Imaging spectroscopy has a particular focus on the identification and classification of plant covers, and even more on the phenological and biochemical characteristics of plants. The reflectance spectra corresponding to vegetation cover provide evidence of the real state of the vegetation cover since they contain information inherent to:

the absorption bands of chlorophyll in the visible region of the spectrum, the high reflectances of healthy vegetation in the near infrared range, and the absorption effects in the mid-infrared region due to water in saturated vegetation, among others (Gates, 1965). According to Clevers *et al.* (2002), the transition band between the visible (red) and near-infrared range of vegetation spectral signatures, known as the red edge region, is where the most important absorption characteristics of vegetation reflectance curves are found, because of the strong contrast or shift between the red and near-infrared range, characterized by an extremely low reflectance value in the visible red, followed by high reflectances in the near infrared range, which is associated with the low red light reflectance of chlorophyll, internal structure and leaf water content, showing that this region of the spectrum is one of the most important.

This absorption characteristic has a width of approximately 100 nm, between 680 and 780 nm, and its inflection point or maximum slope of the curve, known as the red edge position (REP), is generally considered as a parameter for making comparisons between spectral signatures of different plant species, or as an indicator of plant stress and senescence among the same species (Clevers *et al.*, 2002).

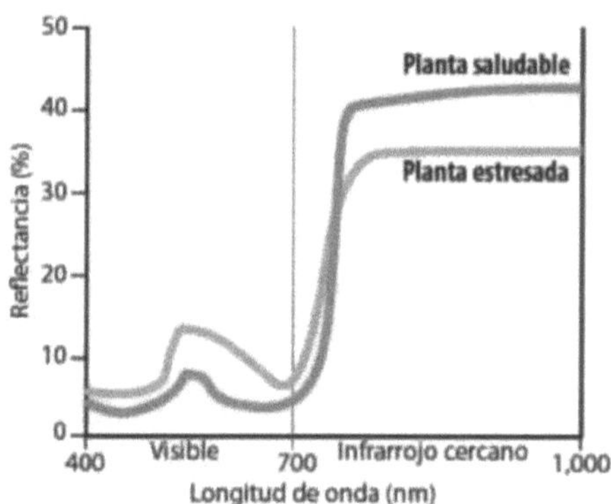

Figure 2. Examples of reflectance spectra of a healthy plant and a stressed plant.
Fluorescence

Fluorescence is a particular type of luminescence, which characterizes substances that are able to absorb energy in the form of electromagnetic radiation and then emit part of that energy in the form of electromagnetic radiation of a different wavelength (Skoog *et al.*, 2007).

At present, chlorophyll fluorescence is a useful tool as a non-invasive method to monitor the status of plants. In addition to being a non-destructive methodology, this technique has the advantages of being fast and highly sensitive (Tremblay *et al.*, 2012).

Measurements in vegetation cover by means of optical sensors

Some reflectance and fluorescence meters are used at the canopy level rather than at the leaf level like foliar chlorophyll meters and therefore can be mounted on tractors and, with the appropriate software and hardware, allow variable fertilizer application based on canopy reflectance (Scharf and Lory, 2009). Reflectance measurements are related to canopy properties such as leaf area index, leaf N and chlorophyll content and biomass (Lemaire *et al.*, 2008; Jongschaap, 2006). Some of the most commonly used instruments are the CropScan, GreenSeeker, CropSpec, Crop Circle and Multiplex. For the application of these measurements to nitrogen fertilization, it is highly recommended to have a reference zone in which the crop is not N deficient, so that the measurements in each zone can be compared with those obtained in the non-deficit zone (Scharf and Lory, 2009). The reflectance data measured by the sensors allow the user to calculate classical vegetation indices such as NDVI (Rouse *et al.*, 1973) and other indices such as the REDVI index (Cao *et al.*, 2013). Barker and Sawyer (2010) have highlighted the importance of developing algorithms based on reflectance measurements for the use of these sensors as a tool for N management in agricultural production, but warn that these algorithms are limited to crop conditions similar to those used in their development.

A recent study on the economic feasibility of applying such measures in corn cultivation in the USA showed that in most of the fields studied there was an economic benefit (Roberts *et al.*, 2010).

Techniques based on reflectance measurements are being widely used in various crops, however there are hardly any application studies for nitrogen fertilizer management in horticultural crops (El-Shikha *et al., 2007;* Pena *et al.*, 2012).

Measuring instruments

SPAD meter

The SPAD-502 chlorophyll meter (Spectrum Technologies, Inc, IL, USA) is a handheld spectrophotometer used to measure the green color of leaves to determine the N nutritional status of plants. Leaf green color is closely related to chlorophyll, which in turn is related to N in leaf. The SPAD meter measures the difference in light transmitted by the leaf at 650 nm, the absorption peak of chlorophyll, and 940 nm, a reference value in the infrared region that depends only on the leaf structure. The SPAD value, calculated by the instrument, is proportional to the relative optical density between the two wavelengths.

Several factors such as the time of measurement, irradiance and the Wdromic status of the plant must be considered when using the SPAD meter (Hoel and Solhaug, 1998; Martinez and Guiamet, 2004). Some works show that the chlorophyll meter is only able to detect severe nitrogen deficiencies (Villeneuve *et al.*, 2002), or requires a well-fertilized plot as a reference (Westcott and Wraith, 1995). Other authors question its use to reliably assess plant N concentration (Himelrick *et al.*, 1993).

Dualex meter®

The Dualex meter (Force-A, Pans, France) is a handheld device for measuring chlorophyll and polyphenols in leaves. Chlorophyll content is estimated, like SPAD, by means of the transmittance ratio in the leaf of two wavelengths in the red and infrared bands of the spectrum. The polyphenol content is estimated from the ratio of chlorophyll fluorescence in the infrared, excited in the red and ultraviolet bands of the spectrum (Goulas *et al.*, 2004).

Dualex provides an NBI index as a ratio between chlorophyll and flavonoid content. This index introduces flavonoid content as a stress factor and thus amplifies possible nutritional deficiencies of the plant. Accumulation of phenolic compounds has been described under nutritional stress (Kiraly, 1964; McClure, 1977; Chishaki and Horiguchi, 1997; Cartelat *et al.*, 2005; Cerovic *et al.*, 2012) and also under imdic stress (Estiarte *et al.*, 1994).

Multiplex Meter®

Multiplex® (Force-A, Orsay, France) is a handheld multi-parametric optical sensor that generates fluorescence in plant tissues using multiple light sources in the ultraviolet, blue, green and red bands. It can simultaneously and non-destructively measure the content of various compounds such as anthocyanins, flavonoids or chlorophyll. There have been studies that demonstrate the relationship between plant nutritional status and Multiplex measurements, such as the studies of Zhang *et al.* (2012) on maize; and Agati *et al.* (2013 and 2015) on plant canopies.

There are hardly any studies with Multiplex in relation to N nutrition of horticultural crops. In studies carried out on corn, comparing it with other sensors such as SPAD and Dualex, significant correlations have been obtained between the values of the SFR index of Multiplex, related to chlorophyll content, and the values of chlorophyll measurements carried out with SPAD and Dualex equipment (Zhang *et al.*, 2012).

Crop-Circle meter

The Crop-Circle ACS-430 is an active light sensor, independent of natural light conditions, which emits radiation at three wavelengths and measures reflectance at the same three wavelengths simultaneously: 670 nm, 730 nm and 780 nm (NIR). In addition to reflectance values, it generates NDVI and NDRE indices. The former is an estimator of vegetation cover (Rouse *et al.*, 1973) and the latter estimates nitrogen content (Barnes *et al.*, 2000).

The sensor can be mounted on almost any type of vehicle and, equipped with the appropriate software, it allows very intensive crop sampling.

There are hardly any studies using Crop-Circle ACS 430 for nitrogen determination in horticultural crops. Pena *et al.* (2012) using the Crop-Circle 470 found significant differences in the ratio of NIR/Green reflectance in melon and Shaver *et al.* (2007) using the NDVI index, observed significant differences in NIR/Green reflectance in melon and Shaver *et al.* (2007) using the NDVI index, observed significant differences in NIR/Green reflectance in melon.

significant in ma^z for different values of nitrogen fertilization, obtaining higher values in the treatments with a higher fertilization rate.

1.2.3. Simulation models

Some authors have proposed the use of models as diagnostic tools for water or nutrient deficit situations (Batchelor *et al.*, 2002; Jones *et al.*, 2003). However, simulation models require a considerable amount of data that are sometimes difficult to obtain. Nevertheless, it has been proposed that some measurements of plants throughout their development can be integrated or "assimilated" into a model in order to correct some of its imperfections (Jongschaap, 2006). This approach is becoming more relevant as tools are being developed to estimate with some ease some data on the nutritional status of crops at any given time (Prevot *et al.*, 2003; Baret *et al.*, 2007). The models, once recalibrated from the measurements obtained at certain points in time, can be used to determine the most appropriate nitrogen fertilization (Houles, 2004).

There are many simulation models of N dynamics in agricultural systems but most of them are for extensive crops, mainly cereals. In general, their objective is to predict the movement and transformation of N in the soil-plant system, but some of them have lower data requirements, which makes them better candidates to be used for nitrogen fertilizer recommendations. Among the most recent or most widely used and tested models are: DSSAT (Jones *et al.*, 2003), STICS (Brisson *et al.*, 2003), APSIM (Keating *et al.*, 2003), NDICEA (Van der Burgt *et al.*, 2006), NLEAP (Shaffer *et al.*, 1991; Shaffer *et al.*, 2010), N_ABLE (Greenwood, 2001), EU-Rotate_N (Rahn *et al.*, 2010a), SMCR_N (Zhang *et al.*, 2010). Information on other N models in the agricultural context can be found in Kersebaum *et al.* (2007) and Cannavo *et al.* (2008).

Of these models, the ones that have been developed with the objective of being used especially in horticultural crops are the N_ABLE, EU-Rotate_N and SMCR_N. These three models have some points in common since the growth modulus follows the pattern applied by Greenwood in the N_ABLE (Greenwood, 2001). The NLEAP model has also been evaluated in some horticultural crops with satisfactory results (Delgado *et al.*, 2000). The EU-Rotate_N model (Rahn *et al.*, 2010b) has been evaluated in other works with acceptable results (Doltra and Munoz, 2010; Doltra *et al.*, 2010).

In general, all models require calibration and validation in a particular environment, defined by crop, soil, climate and cultivation practices. This requires a data set of nitrogen fertilization trials, bearing in mind that the larger the data base on which to calibrate a model, the greater the reliability of the predictions in general.

1.2.4. Systems based on nitrogen balance

These systems can be considered as very simplified models in which the different components of the N balance are estimated or calculated in an approximate way.

The general nitrogen balance equation can be expressed (Meisinger and Randall, 1991) as follows:

$$(N_f + N_r + N_{mn} + N_{fd} + N_{da} + N_a) - (N_l + N_p + N_d + N_v) = \Delta N_s$$

where N_f is the N supplied as fertilizer, N_r is the N supplied in the irrigation water, N_{mn} is the N supplied by net mineralization of soil organic matter, N_{fd} is the N supplied by biological fixation, N_{da} is the N supplied by atmospheric deposition, N_a is the ammonium released from the interlaminar space of the clays, N_l is the N leached, N_p is the N extracted by the crop, N_d is the N losses by denitrification, N_v is the N losses by volatilization and ΔN_s is the variation of the mineral nitrogen content in the soil profile during cultivation.

Examples of such programs are AZODYN (Jeuffroy and Recous, 1999) used in wheat or AZOFERT (Dubrulle *et al.*, 2003) evaluated in cereals and sugar beet, both in France.

- Mineralization

In systems based on soil measurements, simulation models or N balance, the N input by mineralization of soil organic matter can be an important part of the N available to crops and, therefore, its determination is of interest. Fink and Scharpf (2000) and Tremblay et *al.* (2001) in work using horticultural plants estimated a mineralization rate of 5 kgN/ha and week, which implies that in a 90-day cauliflower crop, an N supply of 60 kgN/ha can be achieved.

Measurements of mineralization by incubation tests in the laboratory present problems of transfer to field conditions (Lidon *et al.*, 2005). Therefore, methods have been developed to determine the N mineralization rate in the field (Hatch *et al.*, 2000), which have in common the isolation of a certain amount of soil during an incubation period, thus avoiding processes that can affect the inorganic N pool, such as root uptake, leaching losses and atmospheric deposition. One of these methods is the IER-Core method, proposed by DiStefano and Gholz (1986) and modified by Fisk and Schmidt (1995). In a recently published meta-analysis of different methods for predicting mineralization by soil analysis (Ros *et al.*, 2011) the authors advocate not a single analysis or test but an approach that considers both chemical analysis and other soil properties and environmental conditions.

1.3. Current situation

In the present work, different plant nitrogen measurements have been evaluated for use as a recommendation system for nitrogen fertilization in cauliflower crops. It is included in the national project "Integration of soil and plant measurements and simulation models for the efficient management of nitrogen in horticultural crops" (RTA-2011-00136-C04-02) financed by the I.N.I.A. and in which different Autonomous Communities have collaborated.

The project has two clearly differentiated aspects. The first one aims to evaluate and integrate different methods for the determination of nitrogen fertilization in some horticultural crops. Some based on soil measurements (mineral N content) and others based on plant measurements (nitrate in sap, chlorophyll and canopy reflectance) and simulation models. In the second aspect, the aim is to develop a fertilizer recommendation system in which, based on information easily available to the farmer (soil type, fertilizer data and production of the previous crop, and expected production of the crop in question), a recommendation of nitrogen fertilizer adjusted to the needs can be made. This second aspect is more immediately applicable to the horticultural sector. The work of this project in the Autonomous Community of La Rioja has focused on the cauliflower crop.

2. OBJECTIVES

The general objective of this Doctoral Thesis has been to evaluate different measures of nitrogen in the plant for use as a recommendation system for nitrogen fertilization in cauliflower cultivation.

In turn, to achieve this goal, it has been proposed:

- To study the effect of available nitrogen on production and nitrogen use efficiency in a cauliflower crop (*Brassica oleracea* var. *botrytis*).

- To evaluate the measurement of nitrate concentration in sap and the Nmin method in a cauliflower crop (*Brassica oleracea* var. *botrytis*) to determine the nutritional status.

- To evaluate the reflectance, fluorescence and transmittance of leaves using optical sensors as estimators of nitrogen nutritional status in a cauliflower crop (*Brassica oleracea* var. *botrytis*).

3 MATERIALS AND METHODS

The present chapter on materials and methods has been structured in three main sections:

The first one refers to the nitrogen fertilization trials of cauliflower of the Barcelona variety carried out in the experimental plots of the Finca Valdegon of the CIDA-SIDTA (Servicio de Investigacion y Desarrollo Tecnologico Agroalimentario) of the Government of La Rioja. The second section describes the nitrogen fertilization trials of cauliflower of the Typical variety which were carried out in the same experimental farm. Finally, the third section refers to the nitrogen fertilization trials of cauliflower (var. Casper) carried out at the Experimental Farm of INTIA (Instituto Navarro de Tecnolog^as e Infraestructuras Agroalimentarias) in Sartaguda (Navarra). The methodology used was the same in all the trials, with some exceptions that will be conveniently described.

The laboratory analyses necessary for the study were carried out at CIDA-SIDTA and at the Regional Laboratory of the Government of La Rioja.

2.1. Nitrogen fertilization trials of cauliflower of the Barcelona variety.

Three trials were carried out in experimental plots at Finca Valdegon, in Agoncillo (La Rioja), located at an altitude of 342 m above sea level (UTM, 558.332/4.702.004). Due to the fact that the trials were carried out in different plots and with different fertilizer doses, the description of the three experiments carried out in 2012, 2013 and 2014 has been separated.

2.1.1. Year 2012

Location and initial conditions

In the 2012 experience, cauliflower of the Barcelona variety, with a short cycle of 90 days and destined for the fresh market, was transplanted in a loam-textured soil, classified as *oxyacuic torripsamments* (Soil Survey Staff, 2006). Its properties are presented in Table 1.

Table 1. Physico-chemical properties of the soil in var. Barcelona (2012).

Dep. cm	Sand	Silt %.[1]	Clay[1] % Clay	M.O. %[2]	pH[3]	E.C.[4] dS/m	P ppm[5]	K ppm[5]	Texture[1]
0-15	28,1	49,1	22,8	1,48	8,3	0,31	6,16	194,3	Franco
15-30	26,6	50,2	23,2	1,45	8,3	0,30	7,05	200,9	Loamy loam
30-60	26,6	50,7	22,7	1,14	8,2	0,86	3,48	148,2	Loamy loam
60-90	31,2	44,5	24,4	0,77	8,3	0,88	0,78	113,7	Franco

1) USDA. 2) Oxidizable organic matter. 3) H_2O (1:5). 4) 25°C (1:5). 5) Mehlich III.

2)

Experimental design

Four treatments with four replications were carried out in a randomized design according to the initial Nmin. The elementary plot had an area of 81 m2 and contained 6 crop rows. A maximum rooting depth of 0.6 m was considered for the calculation of nitrogen balance, leachates, etc. (Figure 3).

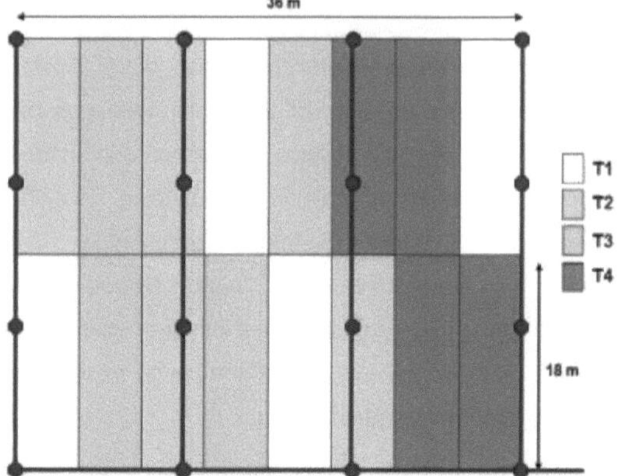

Figura 3. Diseño experimental y distribución de los tratamientos en función del nitrógeno available mineral (Nmin+Nfertilizer) in the 2012 experiment with the Barcelona variety. Available N (kgN/ha): T1: 93; T2: 189; T3: 270; T4: 322. The blue lines and circles indicate the sprinkler network.

As a pre-planting fertilizer, 0-150-275 kg/ha of a complex N-P-K fertilizer was applied and a single application of nitrogen fertilizer was made as a cover crop 31 days after transplanting (DDT) in the form of ammonium nitrosulfate 26-0-0, at a variable dose depending on the available nitrogen of the treatments (Table 2).

Experimental treatments as a function of available N in var. Barcelona (2012).

Treatments	T1	T2	T3	T4
Navailable (kgN/ha)	93	189	270	322
Nmin[1] (kgN/ha)	93	102	121	128
Nfertilizer (kgN/ha)	0	87	149	194

1) Nitrate and Ammonium

Transplanting was carried out on August 13 on plateaus separated by 1.5 m between axes, with two planting beds per plateau (0.75 x 0.6 m), resulting in a density of 23,412 plants/ha. Irrigation was carried out by sprinkler irrigation according to FAO56 dual program (Allen *et al.,* 1998). A daily water balance up to a depth of 0.6 m was carried out with data supplied by the SIAR (Agroclimatic Information Service of La Rioja) of the Agroclimatic Station located on the same farm. Crop cover and height data were entered daily using the models described below. To check the evolution of the wetted profile, eight Watermark probes were installed in four stations distributed throughout the trial at two different depths, 0.3 and 0.6 m. The phytosanitary treatments were as described below. The phytosanitary treatments were those recommended to maintain the good sanitary status of the crop.

Sampling and determinations

To have data with which to update the daily water balance, data on crop height and cover were taken for the four treatments. For the determination of cover, digital photographs were taken periodically of all the replicates of the trial and the crop cover was calculated using the image processing program Gimp® (Campillo *et al.*, 2010). On the same day, the height of five plants per replicate was measured.

$$COB = \frac{A}{1 + EXP\left(- B\left(IT_{5.5} - C\right)\right)} \qquad [1]$$

Crop cover (COB, %) was modeled as a function of the thermal integral with a threshold of 5.5°C (IT5.5) (Maroto, 2002) according to a logistic model of the type:

Where A, estimates the maximum crop coverage, B estimates the initial slope and C estimates the time at which 50% of the final coverage is reached.

$$ALT = A + \frac{B}{1 + EXP\left(- C\left(IT_{5.5} - D\right)\right)} \qquad [2]$$

Crop height (m) was treated similarly, as a function of the thermal integral with a threshold of 5.5°C according to a logistic model of the type:

Where A, estimates the initial height of the crop at transplanting and A+B the maximum height of the crop. Parameters C and D have the same meaning as indicated in the coverage model.

The adjustment of both models was performed by nonlinear regression and the subsequent comparison between treatments of the corresponding parameters by Student's *t-test*.

Nmin (nitrate and ammonium) was determined at planting, forty-eight days after transplanting (DDT) and at harvest. For this purpose, two samples were taken at each elemental plot, mixing the soil by layers, at 0-15, 15-30, 30-60 and 60-90 cm depth. The soil was extracted with 1M KCl and analyzed for nitrate and ammonium content with the AxFlow AA3 autoanalyzer.

The rate of mineralization was determined approximately every fifteen days. For this purpose, 0.25 m long PVC tubes were used with a resin filter installed at the distal end to collect the leached nitrate (DiStefano and Gholz, 1986). One tube was inserted in each elementary plot, keeping the soil undisturbed. The net mineralization in each tube was calculated as the difference between the final and initial Nmin content plus the nitrogen retained in the resins.

At 28, 49 and 63 DDT and at harvest, fresh weight, dry weight, total N and $N\text{-}NO_3^-$ were determined in leaves and pellets of five plants per elemental plot. Total N was analyzed by the Kjeldhal method (AOAC, 1990) and nitrate content was analyzed by extraction with 0.025 M Ah$(SO_4)_3H_8H_2O$ and nitrate concentration was measured by the ion-selective electrode method (Jones and Case, 1990; Miller, 1998).

To all the results obtained for dry weight and total nitrogen content in the trials for each variety, the general nitrogen model of Greenwood (1986), the model for the *Brassica* genus of Greenwood (1996) and the EU_Rotate model (Rahn *et al.*, 2010a and 2010b) were applied in order to check which model best discriminated the nitrogen deficient treatments from the non-deficient ones. Deficient treatments were considered as those that showed a significantly lower production than the rest and whose available nitrogen was below the levels considered as recommended in the literature and in the experimental design of the trials. Figure 4 shows the values obtained with the Barcelona variety in the 2012, 2013 and 2014 trials, for biomasses greater than 1 Mg/ha, as well as the models mentioned above. Biomass values lower than 1 Mg/ha have not been considered since for these biomass values the concentration of cntic nitrogen is independent of aerial biomass (Justes *et al.*, 1994). Of the three models
studied, the general model of Greenwood (1986) correctly located 97.4% of the treatments considered as non-deficit and 62.1% of the deficit treatments. The Greenwood (1996) and EU-Rotate (Rahn *et al.*, 2010a and 2010b) models correctly placed 100% of the deficit treatments but only 18% and 39% of the non-deficit treatments, respectively. Therefore, the general form of this model will be used in the analysis of results.

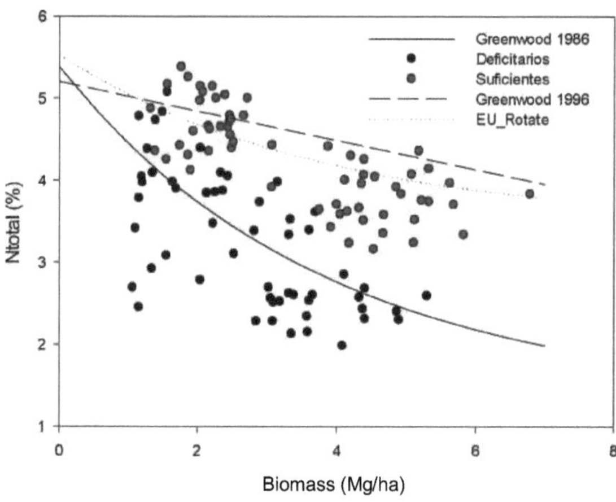

Figure 4. Total leaf nitrogen (%) and biomass (Mg/ha) of cauliflower var. Barcelona plants in the 2012,

2013 and 2014 trials. Lines corresponding to Greenwood (1986), *Brassica* (Greenwood (1996) and EU-Rotate (Rahn *et al.*, 2010a and 2010b) general nitrogen models. Deficient: treatments and replicates whose available nitrogen is below the levels considered as recommended in the literature and in the experimental design itself. Biomass values higher than 1 Mg/ha.

Harvesting took place between 66 and 84 DDT. In a subplot of 9 m^2 all plants were collected. The total weight of each plant and the weight of the pellet with and without the basal leaves were determined. All pellets of Extra or First quality and a diameter greater than 11 cm were considered commercial (EEC, 1998). In this and subsequent trials, harvest data were adjusted for available mineral nitrogen in a two-stage linear regression (Equation 3), seeking the highest yield value for the lowest value of available mineral nitrogen in the soil.

$$y = b * Ndisp(Ndisp < a) + b * a(Ndisp \geq a) \qquad [3]$$

Where y is the pellet yield, *Ndisp* is the available mineral nitrogen content in the soil, and a and b are parameters of the regression. The parameter a corresponds to the value of the abscissa from which the slope becomes zero. The product $a \times b$ is the value of the ordinate that estimates the maximum value of the yield.

On four dates throughout the crop, and on the same leaf, measurements were made using SPAD (Mod. 502, Minolta), Dualex® and Multiplex® equipment. 25 (Force-A). Measurements were made at 11:00 h, on one fully expanded leaf per plant, on ten plants per elementary plot. Five pedoles of these plants were chopped and mixed in the laboratory to extract the sap by pressing, and a quota of this sample was diluted (when necessary) in deionized water and the N-NO3 content⁻ was measured using a portable reflectometer RQflex©.

The SPAD sensor is a meter of relative chlorophyll concentration by means of light transmitted through the leaf at 650 nm (photosynthetically active wavelength) and 940 nm.

The Dualex sensor provides three indices. The Chl index is an estimator of chlorophyll content by means of the ratio of transmittances in the leaf between the red at 710 nm and the near infrared (NIR) at 850 nm. The FLAV index is an estimator of leaf flavonoid content that increases under nitrogen deficiency conditions. The nitrogen balance index, NBI, is calculated as the ratio between Chl and FLAV.

The Chl index is an estimator of chlorophyll content and is expressed as:

$$Chl = \frac{\tau_{NIR} - \tau_{R}}{\tau_{R}} \qquad [4]$$

where τ_{NIR} and τ_{R} are the transmittances in the near infrared band (NIR) and in the red band (R).

The FLAV index is an estimator of the flavonoid content and is expressed as follows

$$FLAV = \log \frac{CF_{NIR}^{R}}{CF_{NIR}^{UV-A}} \qquad [5]$$

where CFNIR is the fluorescence of chlorophyll in the near infrared (NIR) excited by red (R) and ultraviolet (UV-A). The difference between the two fluorescences is proportional to the amount of flavonoids present in the leaf epidermis. From the Chl and FLAV indices it is possible to calculate the NBI index as the quotient between them.

The Multiplex sensor generates fluorescence in plant tissues using the excitation of various light sources to obtain up to twenty parameters related to the physiological state of the plant. In this work we have used the SFR-R index, which is an estimator of chlorophyll content obtained by the ratio of fluorescence between red and far red, and the NBI.

The SFR index is expressed as:

$$SFR\text{-}R = \frac{F_{NIR}^{visible}}{F_{R}^{visible}} \qquad [6]$$

where FNIR and FR are the near-infrared (NIR) and red (R) fluorescence of chlorophyll, respectively, both excited by the visible.

The FLAV index is obtained in a similar way as described for the DUALEX sensor. From the SFR-R and FLAV indices it is possible to calculate the NBI index as the chlorophyll/flavonol ratio.

On October 1, at 49 DDT, a reflectance measurement was made with the Crop-Circle ACS-430 (Holland Scientific) which measures soil and crop reflectance at 670, 730 and 780 nm. The Crop-Circle sensor is an active sensor that emits radiation in the visible band and therefore does not depend on solar radiation for reflectance measurement. Two 5 m paths were made on two interior lines of each elementary plot, at a height of 0.9 m above the ground, which means a sweep of 0.75 m at ground surface level. This sensor provides the NDVI and NDRE indices. The NDVI index (Rouse *et al.*, 1973) is one of the most widely used vegetation indices, ranging from 0.1 for bare soil to 0.9-1.0 in fully developed vegetation cover. It is calculated as the quotient of the difference and the sum of the NIR (780 nm) and red (670 nm) reflectances. The NDRE index is similar to NDVI but uses reflectance at 730 nm instead of 670 nm and is sensitive to changes in leaf chlorophyll A or nitrogen content. The lower value of the NDRE index means an increase in reflectance at 730 nm indicating a lower value of chlorophyll content (Fitzgerald *et al.*, 2006).

A nitrogen balance in the soil-plant system was elaborated. In this balance, the following were established as input parameters to the system: the initial mineral nitrogen present in the soil, that applied in the form of fertilizer and the nitrogen supplied through the mineralization of soil organic matter during the crop cycle, and as output parameters: the mineral nitrogen present in the soil at the end of the crop, the nitrogen extracted by the plant and the nitrogen leached.

$$(N_{min} + N_{fert} + N_{miner}\ N_{minf} - N_{ext} - N_{lix}) = \Delta N_s \tag{7}$$

Where N_{min} is the initial mineral nitrogen present in the soil, N_{fert} is the N supplied as fertilizer, N_{miner} is the N supplied by the net mineralization of the soil organic matter, N_{lix} is the N leached, N_{ext} is the N extracted by the crop, N_{minf} is the N extracted by the soil, N_{minf} is the N leached from the soil, N_{miner} is the N extracted by the crop and N_{minf} is the N extracted from the soil by the crop. 27 is the mineral nitrogen in the soil at the end of cultivation, and ΔN_s is the variation of the mineral nitrogen content in the soil profile during cultivation.

The $N\text{-}NO_3$ leachate⁻ was calculated from a water balance and the average $N\text{-}NO_3$ concentration⁻ between the 30-60 and 60-90 cm layers. Nitrogen use efficiency (NUE) was calculated as the ratio of commercial yield to available nitrogen (initial mineral N + applied N) (Moll *et al.*, 1982).

In the statistical analysis of the results, analysis of variance (ANOVA), linear and nonlinear regression analysis, and Student's and Tukey's tests were used using the SYSTAT® 12 program. The normality and homoscedasticity of the data were checked prior to the analyses. Table 3 shows a summary of some phenological and crop data.

Data from the trial in var. Barcelona (2012).

Date of plantation	13/08/2012
Planting density (plants.ha-1)	23.412
Number of irrigations	16
Cumulative irrigation (mm)	211
Precipitation (mm)	121
Subscriber	13/09/2012
Start of pellet formation*.	01/10/2012
Collection Penodo	17/10/12-05/11/12

*The beginning of pellet formation is considered to be the moment when pellets of 1 mm in diameter are detected.

2.1.2. Year 2013

Localization and initial conditions

The experiment was carried out at the SIDTA farm. Cauliflower of the Barcelona variety was used in a soil with a sandy loam texture and classified as *oxyacuic torriorthens* (Soil Survey Staff, 2006) and its properties are presented in Table 4.

Table 4. Physico-chemical properties of the soil in var. Barcelona (2013).

Dep. cm	Sand	Silt %.[1]	Clay[1] % Clay	M.O. %[2]	pH[3]	E.C.[4] dS/m	P ppm[5]	K ppm[5]	Texture[1]
0-15	57,7	32,1	10,3	0,56	8,4	0,16	4,2	132,0	Fco. arenoso
15-30	57,0	32,6	10,5	0,56	8,5	0,16	7,6	111,9	Fco. arenoso
30-60	73,9	19,1	7,0	0,26	8,7	0,10	0,8	50,0	Fco. arenoso
60-90	84,3	11,4	4,3	0,13	8,8	0,09	0,1	27,9	Fco. arenoso

1) USDA. 2) Oxidizable organic matter. 3) H_2O (1:5). 4) 25°C (1:5). 5) Mehlich III.

Experimental design

The characteristics of the trial in terms of field distribution, irrigation, calculation of water requirements, crop cover and height, and installation of Watermark probes were similar to those of 2012. It should be noted that on September 6 there was a hail event that affected the crop, delaying its development, although it did not affect the final development of the plants.

Transplanting was carried out on August 7 on plateaus separated by 1.5 m between axes, with a double row per plateau and a density of 20,550 plants/ha. In pre-planting, 0-140-260 kg/ha of a complex N-P-K fertilizer was applied, and a nitrogen fertilizer application was made 26 days after transplanting (DDT) in the form of ammonium nitrosulfate 26-0-0, at a variable dose depending on the available nitrogen (Table 5).

Experimental treatments as a function of available N in var. Barcelona (2013).

Treatments	T1	T2	T3	T4
Navailable (kgN/ha)	67	130	193	260
Initial Nmin (kgN/ha)	67	80	93	130
Nfertilizer (kgN/ha)	0	50	100	130

The design was established according to the initial soil mineral nitrogen (Nmin) and consisted of four treatments with four replications (Table 5). The elementary plot had an area of 81 m² and contained 6 Kneas of crop (Figure 5).

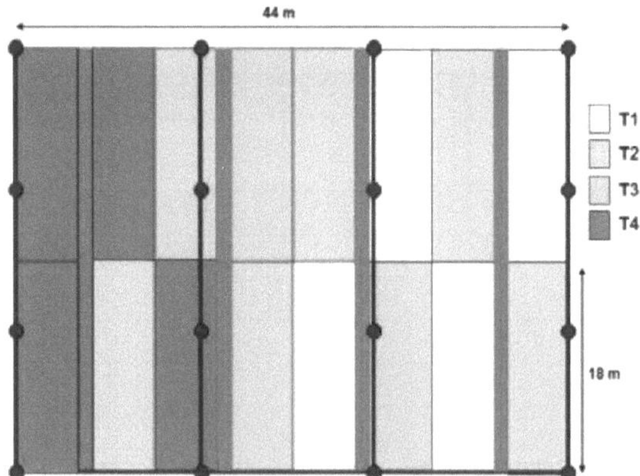

Experimental design and distribution of available mineral nitrogen treatments in the 2013 experimental plot with the Barcelona variety. Available N (kgN/ha): T1: 67; T2: 130; T3: 193; T4: 260. The blue lines and circles indicate the sprinkler network. Green areas indicate corridors.

Sampling and determinations

Nmin (nitrate and ammonium) was determined at planting, forty-eight days after transplanting (DDT) and at harvest. For this purpose, two samples were taken in each elemental plot at 0-15, 15-30, 30-60 and 60-90 cm depth, mixing the soil in layers.

The rate of mineralization was determined approximately every fifteen days using resin filters (DiStefano and Gholz, 1986). One tube was inserted in each elemental plot of the T1 and T4 treatments, keeping the soil undisturbed.

At 23, 43, 56 and 76 DDT and at harvest, fresh weight, dry weight and total N were determined in leaves and pellets of five plants per elemental plot.

Harvesting was carried out between 64 and 86 DDT in a 9 m subplot^2 where all plants were collected. Total weight and pellet weight were determined for each plant. All pellets of Extra or First quality and a diameter greater than 11 cm were considered as commercial (EEC, 1998).

On five dates during cultivation, leaf measurements were made with SPAD (Mod. 502, Minolta), Dualex® and Multiplex® (Force-A) equipment, at 11:00 h, on one adult leaf per plant, on ten plants per elementary plot. Five petioles of these leaves were chopped and mixed in the laboratory to extract the sap and determine the nitrate content by the method explained in the 2012 trial. On three dates, reflectance measurements were taken at 670, 730 and 780 nm, with the Crop-Circle™ ACS-430 (Holland Scientific). A 12-meter path was taken over the two inner lines of each elementary plot.

As in the previous year, a nitrogen balance is performed. The $N\text{-}NO_3$ leachate⁻ was calculated from a water balance and the average $N\text{-}NO_3$ concentration⁻ between the 30-60 and 60-90 cm layers. Nitrogen use efficiency (NUE) was calculated as the ratio of commercial yield to available nitrogen (initial mineral N + applied N) (Moll *et al.*, 1982).

The statistical treatment of the results has been similar to that described in the 2012 trial. A summary of some phenological and crop data is presented in Table 6.

Data from the trial in var. Barcelona (2013).

Date of plantation	07/08/2013
Planting density (plants.ha-1)	20.550
Number of irrigations	20
Cumulative irrigation (mm)	194
Precipitation (mm)	65
Subscriber	02/09/2013
Start of pellet formation*.	30/09/2013
Harvesting period	09/10/13-31/10/13

*The beginning of pellet formation is considered to be the moment when pellets of 1mm diameter are detected.

2.1.3. Year 2014

Localization and initial conditions

The experiment was carried out at the SIDTA farm. Cauliflower of the Barcelona variety was used in a loam-textured soil and was classified as *oxyacuic torriorthens* (Soil Survey Staff, 2006) and its properties are presented in Table 7.

Table 7. Soil physicochemical properties in var. Barcelona (2014).

Dep. cm	Sand	Silt %.[1]	Clay[1] % Clay	M.O. %[2]	pH[3]	E.C.[4] dS/m	P ppm[5]	K ppm[5]	Texture[1]
0-15	28,10	49,11	22,79	1,48	8,26	0,31	6,2	194,3	Franco
15-30	26,61	50,19	23,20	1,45	8,30	0,30	7,0	200,9	Franco
30-60	26,60	50,70	22,71	1,14	8,22	0,86	3,5	148,1	Franco
60-90	31,15	44,48	24,37	0,77	8,29	0,88	0,8	113,7	Franco

1) USDA. 2) Oxidizable organic matter. 3) H_2O (1:5). 4) 25°C (1:5). 5) Mehlich III.

Experimental design

The characteristics of the trial in terms of field distribution, irrigation, calculation of water requirements, crop cover and height, and installation of Watermark probes were similar to those of 2012 and 2013. Transplanting was carried out on August 11 on plateaus separated by 1.5 m between axes, with a double row per plateau and a density of 20,440 plants/ha. In pre-planting, 0-82-213 kg/ha of a complex N-P-K fertilizer was applied, and a nitrogen fertilizer application was made 26 days after transplanting (DDT) in the form of ammonium nitrosulfate 26-0-0, at a variable dose depending on the available nitrogen (Table 8).

Experimental treatments as a function of available N in var. Barcelona (2014).

Treatments	T1	T2	T3	T4

Navailable (kgN/ha)	0	130	190	260
Initial Nmin (kgN/ha)	72	94	111	124
Nfertilizer (kgN/ha)	0	36	79	136

The block design was established according to the initial soil mineral nitrogen (Nmin) and consisted of four treatments with five replicates in treatments T1 and T2 and four replicates in treatments T3 and T4 (Table 8). The elementary plot had an area of 81 m2 and contained 6 Kneas of crop (Figure 6).

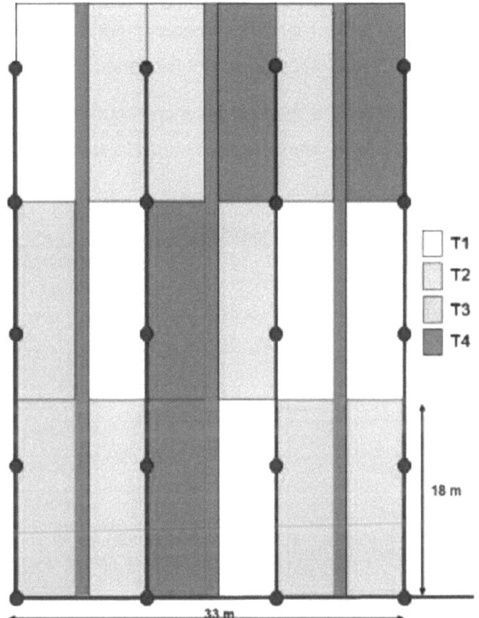

Experimental design and distribution of available mineral nitrogen treatments in the 2014 experimental plot with the Barcelona variety. Available N (kgN/ha): T1: 72; T2: 130; T3: 190; T4: 260. The blue lines and circles indicate the sprinkler network. Green areas indicate aisles.

Sampling and determinations

Nmin (nitrate and ammonium) was determined at planting, fifty-four days after transplanting (DDT) and at harvest. For this purpose, two samples were taken in each elemental plot at 0-15, 15-30, 30-60 and 60-90 cm depth, mixing the soil in layers.

The rate of mineralization was determined approximately every fifteen days in all elementary plots of the trial according to the methodology described for the 2013 trial. One tube was inserted in each elementary plot of all treatments while maintaining the soil undisturbed.

At 29, 55, 74 DDT and at harvest, fresh weight, dry weight and total N were determined in leaves and pellets from five plants per elementary plot. Harvesting was carried out between 71 and 90 DDT in a 9 m subplot[2] where all plants were collected. Total weight and pellet weight were determined for each plant. All pellets of Extra or First quality and a diameter greater than 11 cm were considered as commercial (EEC, 1998).

On three dates during cultivation, leaf measurements were made with SPAD (Mod. 502, Minolta), Dualex® and Multiplex® (Force-A) equipment, at 11:00 h, on one adult leaf per plant, on ten plants per elementary plot. Five petioles of these leaves were chopped and mixed in the laboratory to extract the sap and determine the nitrate content by the method explained in the 2012 trial. On three dates, reflectance measurements were taken at 670, 730 and 780 nm, with the Crop-Circle™ ACS-430 equipment (Holland Scientific) over a 12-meter path.

As in the previous year, a nitrogen balance was performed. The data treatment has been similar to that of previous years. Table 9 shows a summary of some phenological and crop data.

Table 9. Data from the trial in var. Barcelona (2014).

Date of plantation	11/08/2014
Planting density (plants.ha-1)	20.440
Number of irrigations	16
Cumulative irrigation (mm)	224
Precipitation (mm)	73
Subscriber	11/09/2014
Start of pellet formation*.	03/10/2014
Collection Penodo	22/10/14-10/11/14

*The beginning of pellet formation is considered to be the moment when pellets of 1 mm in diameter are detected.

2.2. Nitrogen fertilization trials of cauliflower of the Typical variety.

This chapter describes the cauliflower experiments on Typical cauliflower under field conditions. Due to the fact that the trials were conducted in several plots and subject to different conditions, treatments and fertilizer doses, it has been preferred to separate the description of the two experiments, from the years 2013 and 2014. The trials were carried out in experimental plots of the Valdegon Farm of the Government of La Rioja, whose location has been described in section 3.1.

2.2.1. Year 2013

Localization and initial conditions

The experiment was carried out at the SIDTA farm. Typical cauliflower variety was used, intended for industry, with a long cycle, around 180 days, in a loam-textured soil and classified as *oxyacuic torriorthens* (Soil Survey Staff, 2006) and its properties are presented in Table 10.

Experimental design

The characteristics of the trial in terms of field distribution, irrigation, calculation of water requirements, crop cover and height, and installation of Watermark probes were similar to those of the Barcelona variety.

Table 10. Physico-chemical properties of the soil in var. Typical (2013).

Dep. cm	Sand	Silt %.[1]	Clay[1] % Clay	M.O. %[2]	pH[3]	E.C.[4] dS/m	P ppm[5]	K ppm[5]	Texture[1]
0-15	39,2	45,5	15,3	1,16	8,2	0,39	8,2	155,9	Franco
15-30	50,9	36,7	12,4	1,01	8,3	0,33	10,6	131,7	Franco
30-60	31,7	50,2	18,1	1,03	8,6	0,25	4,5	69,4	Franco
60-90	33,3	52,0	14,7	0,66	8,3	0,98	1,4	32,0	Franco

1) USDA. 2) Oxidizable organic matter. 3) H_2O (1:5). 4) 25°C (1:5). 5) Mehlich III.

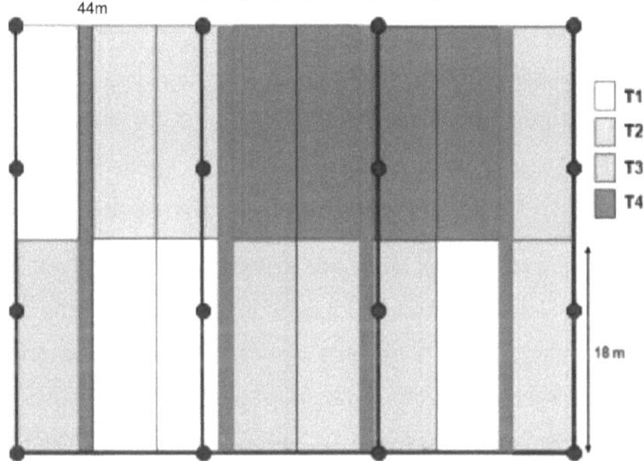

Experimental design and distribution of available mineral nitrogen treatments in the plot of the 2013 experiment with the Typical variety. Available N (kgN/ha): T1: 84; T2: 130; T3: 190; T4: 260. The blue lines and circles indicate the sprinkler network. The areas green indicate corridors.

Transplanting was carried out on August 8 on plateaus separated by 1.5 m between axes, with a double row per plateau and a density of 20,833 plants/ha. Irrigation was by sprinkler. In pre-planting, 0-140-260 kg/ha of a complex N-P-K fertilizer was applied, and two applications of nitrogen fertilizer were made in cover crops, 24 and 48 days after transplanting (DDT) in the form of ammonium nitrosulfate 26-0-0, at a variable dose depending on the available nitrogen (Table 11).

The block design was established according to the initial soil mineral nitrogen (Nmin) and consisted of four treatments with four replications (Table 11). The elementary plot had an area of 81 m^2 and contained 6 crop rows (Figure 7). On September 6, a hail event caused moderate defoliation in all plots, from which the plants subsequently recovered.

Experimental treatments as a function of available N in var. Typical (2013).

Treatments	T1	T2	T3	T4
Navailable (kgN/ha)	84	130	190	260
Initial Nmin (kgN/ha)	84	95	122	175
Nfertilizer (kgN/ha)	0	35	68	85

Sampling and determinations

Nmin (nitrate and ammonium) was determined at planting, forty days after transplanting (DDT) and at harvest. For this purpose, two samples were taken in each elemental plot at 0-15, 15-30, 30-60 and 60-90 cm depth, mixing the soil in layers.

The rate of mineralization was determined approximately every fifteen days. One tube was inserted in each elemental plot of the T1 and T4 treatments, keeping the soil undisturbed.

At 20, 40, 62, 106 DDT and at harvest, fresh weight, dry weight and total N were determined in leaves and pellets from five plants per elementary plot. The harvest was carried out between 106 and 207 DDT in a subplot of 9 m^2 where all the plants were collected. Total weight and pellet weight were determined for each plant. All pellets of Extra or First quality and a diameter greater than 11 cm were considered as commercial (EEC, 1998).

On four dates during cultivation, leaf measurements were made with SPAD (Mod. 502, Minolta), Dualex® and Multiplex® (Force-A) equipment, at 11:00 h, on one adult leaf per plant, on ten plants per elementary plot. Five petioles of these leaves were chopped and mixed in the laboratory for sap extraction. On three dates, reflectance measurements were taken at 670, 730 and 780 nm, with the CropCircle™ ACS-430 (Holland Scientific). A 12-meter path was taken over the two inner lines of each elementary plot.

As in the previous year, a nitrogen balance was performed. The data treatment has been similar to that of previous years. Table 12 shows a summary of some phenological and crop data.

Table 12. Trial data in var. typical (2013).

Date of plantation	08/08/2013
Planting density (plants.ha-1)	20.833
Number of irrigations	24
Cumulative irrigation (mm)	281
Precipitation (mm)	219
First subscriber	02/09/2013
Second subscriber	25/09/2013
Start of pellet formation*.	11/11/2013
Collection Penodo	21/01/14-06/03/14

*The beginning of pellet formation is considered to be the moment when pellets of 1 mm in diameter are detected.

2.2.2. Year 2014

Location and initial conditions

The experiment was carried out at the SIDTA farm. Cauliflower of the long-cycle Typical variety was used in a soil with a sandy loam texture and classified as *oxyacuic torriorthens* (Soil Survey Staff, 2006) and its properties are presented in Table 13.

Table 13. Physico-chemical properties of the soil in var. Typical (2014).

Dep. cm	Sand	Silt %.[1]	Clay[1] % Clay	M.O. %[2]	pH[3]	E.C.[4] dS/m	P ppm[5]	K ppm[5]	Texture[1]
0-15	57,7	32,1	10,3	0,56	8,4	0,16	4,2	132,0	Fco. arenoso
15-30	57,0	32,6	10,5	0,56	8,5	0,16	7,6	111,9	Fco. arenoso
30-60	73,9	19,1	7,0	0,26	8,7	0,10	0,8	50,0	Fco. arenoso
60-90	84,3	11,4	4,3	0,13	8,8	0,09	0,1	27,9	Fco. arenoso

1) USDA. 2) Oxidizable organic matter. 3) H_2O (1:5). 4) 25°C (1:5). 5) Mehlich III.

Experimental design

The characteristics of the trial in terms of field distribution, irrigation, calculation of water requirements, crop cover and height, and installation of Watermark probes were similar to those of the Barcelona variety.

Transplanting was carried out on August 5 on plateaus separated by 1.5 m between axes, with a double row per plateau and a density of 22,222 plants/ha.

Irrigation was by sprinkler. In pre-planting, 0-82-213 kg/ha of a complex N-P-K fertilizer was applied, and in cover crops, two applications of nitrogen fertilizer were made, at 29 and 42 days after transplanting (DDT) in the form of ammonium nitrosulfate 26-0-0, at a variable dose depending on the available nitrogen (Table 14).

Table 14. Experimental treatments as a function of available N in Typical (2014).

Treatments	T1	T2	T3	T4
Navailable (kgN/ha)	95	170	230	300
Initial Nmin (kgN/ha)	95	130	135	177
Nfertilizer (kgN/ha)	0	40	95	123

The block design was established according to the initial soil mineral nitrogen (Nmin) and consisted of four treatments with four replications (Table 14). The elementary plot had an area of 81 m² and contained 6 crop rows (Figure 8). On February 2, 2015, a flooding of the Ebro river caused a flooding in the trial plot, terminating the experiment without having completed the

cosecha.

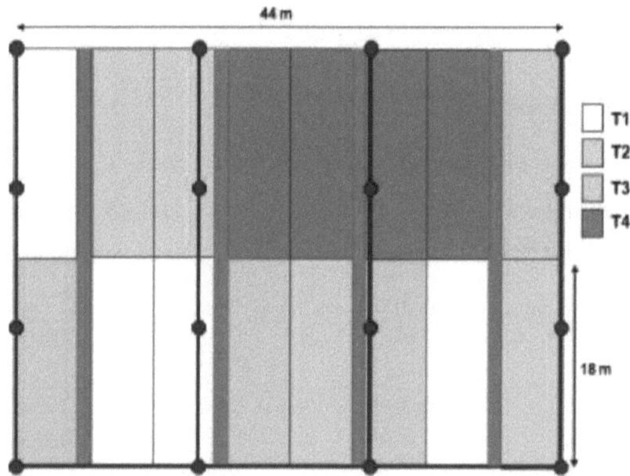

Figura 8. Diseño experimental y distribución de los tratamientos de nitrógeno mineral disponible in the plot of the 2014 experiment with the Typical variety. Navailable (kgN/ha): T1: 95; T2: 170; T3: 230; T4: 300. The blue lines and circles indicate the sprinkler network. Green areas indicate corridors.

Sampling and determinations

The Nmin (nitrate and ammonium) in the plantation was determined at 28 days after transplanting (DDT), at 41 DDT, at 57 DDT and at 83 DDT. For this purpose, two samples were taken in each elemental plot at 0-15, 15-30, 30-60 and 60-90 cm of depth, mixing the soil in layers.

At 27, 41, 57, 79 DDT and at harvest, fresh weight, dry weight and total N were determined in leaves and pellets of five plants per elemental plot.

Due to the flooding caused by the Ebro River flooding, only approximately 30% of the total could be harvested. This partial harvest was carried out between 147 and 176 DDT in a subplot of 9 m^2 where all the plants were collected. The total weight and pellet weight of each plant was determined. All pellets of Extra or First quality and a diameter greater than 11 cm were considered as commercial (EEC, 1998).

On four dates during cultivation, leaf measurements were made with SPAD (Mod. 502, Minolta), Dualex® and Multiplex® (Force-A) equipment, at 11:00 h, on one adult leaf per plant, on ten plants per elementary plot. Five petioles of these leaves were chopped and mixed in the laboratory for sap extraction by pressing. On five dates, reflectance measurements were taken at 670, 730 and 780 nm, with the Crop-Circle™ ACS-430 (Holland Scientific). A 12-meter path was taken along the two inner lines of each elementary plot.

As in the previous year, a nitrogen balance was performed. The data treatment has been similar to that of previous years. Table 15 shows a summary of some phenological and crop data.

Table 15. Trial data in var. typical (2014).

Date of plantation	05/08/2014

Planting density (plants.ha-1)	20.833
Number of irrigations	17
Cumulative irrigation (mm)	233
Precipitation (mm)	204
First subscriber	03/09/2014
Second subscriber	16/09/2014
Start of pellet formation*.	22/10/14
Collection Penodo	30/12/14-28/01/15

*The beginning of pellet formation is considered to be the moment when pellets of 1 mm in diameter are detected.

2.3. Nitrogen fertilization trials of the cauliflower variety Casper.

This chapter describes the cauliflower experiments under field conditions. Because trials were conducted on several plots and subjected to different conditions, treatments and fertilizer rates, we preferred to separate the description of the two experiments in 2012 and 2014.

The two trials were carried out in experimental plots at the INTIA Experimental Farm in Sartaguda (Navarra), which is located 337 meters above sea level (UTM, 577.967; 4.690.730).

2.3.1. Year 2012

Location and initial conditions

A nitrogen fertilization trial was carried out on a short-cycle cauliflower var. Casper crop at the INTIA experimental farm. The soil properties are presented in Table 16.

Table 16. Physico-chemical properties of the soil in var. Casper (2012).

Dep. cm	Sand	Silt %.[1]	Clay[1] % Clay	M.O. %[2]	pH[3]	E.C.[4] dS/m	P ppm[5]	K ppm[5]	Texture[1]
0-30	60,6	28,8	10,6	1,15	8,15	0,81	54,7	300,7	sandy-loam
30-60	58,4	31,2	10,4	0,91	8,24	0,86	36,0	215,3	sandy-loam
60-90	65,7	27,2	7,1	0,61	8,33	0,68	13,3	144,2	sandy-loam

1) USDA. 2) Oxidizable organic matter. 3) H2O (1:2,5). 4) (1:1). 5) Mehlich III.

The Casper variety of cauliflower was used as planting material. Transplanting was carried out on August 2 at a planting density of 22,222 plants/ha, on plateaus separated by 1.50 m between axes and 60 cm between plants, with two growing beds per plateau (Figure 6). ETc was calculated following the FAO dual Kc approach (Allen *et al.*, 1998). Meteorological data and ETo were obtained from the weather station located on the same farm. A sprinkler irrigation system was used.

Experimental design

Four treatments with different levels of available Nmin (initial Nmin + Nfertilizer) and four replicates were differentiated in a block design based on initial Nmin. The elementary plot had an area of 81 m^2 and 6 crop rows (Figure 9). For the determination of mineral, nitrogen, nitric and ammoniacal nitrogen, present in the soil at the beginning of the crop, the depths of 0-15, 15-30, 30-60 and 60-90 cm were sampled. The Nmin values of the different treatments and the fertilizer doses used are shown in Table 17.

The top dressing was made in a single application of nitrogen fertilizer in the form of 26% ammonium nitrosulfate on September 7. In all treatments, 100-150 kg/ha of a P-K complex fertilizer was applied as a background fertilizer.

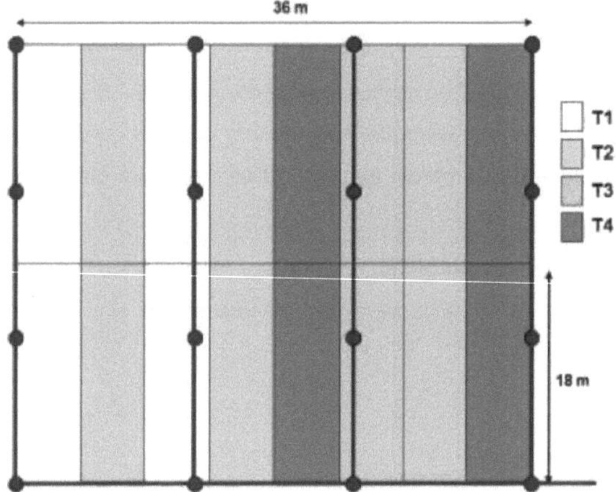

Figura 9. Diseño experimental y distribución de los tratamientos de nitrógeno mineral disponible in the plot of the 2012 experiment with the Casper variety. Navailable (kgN/ha): T1: 259; T2: 359; T3: 432; T4: 524. The blue Kneas and circles indicate the sprinkler network.

Table 17. Experimental treatments as a function of available N in var. Casper (2012).

Treatments	T1	T2	T3	T4
Navailable (kgN/ha)	259	359	432	524
Initial Nmin (kgN/ha)	259	309	432	424
Nfertilizer (kgN/ha)	0	50	0	100

Sampling and determinations

Nmin (nitrate and ammonium) was determined at planting, 62 days after transplanting (DDT) and at harvest. For this purpose, two samples were taken in each elemental plot at 0-15, 15-30, 30-60 cm depth, mixing the soil in layers.

At transplanting, at 20, 35, 64 DDT and at harvest, fresh weight, dry weight and total N in leaves and pellets of five plants per elementary plot were determined. Harvesting was done between 89 and 99 DDT in a 9 m subplot^2 where all plants were collected. Harvesting was staggered, on three dates, 30 October, 6 and 9 November, after a crop cycle of 89 days. The

total production, commercial and average weight of the pellet. The size of the skin in the reflection has been determined by the destination of the harvest, in this case for industry, being usual in this area values between 1,200 and 1,600 g per skin.

On four dates during cultivation, leaf measurements were made with SPAD equipment (Mod. 502, Minolta) at 11:00 h, on one adult leaf per plant, on ten plants per elementary plot. Five petioles of these leaves were chopped and mixed in the laboratory to extract the sap.

Nitrogen use efficiency (NUE) was calculated as the ratio between commercial production and available nitrogen (initial mineral N + applied N) (Moll *et al.*, 1982). As in previous years, a nitrogen balance was performed.

The data treatment has been similar to that of previous years. Table 18 shows a summary of some phenological and crop data.

Table 18. Casper var. trial data (2012).

Date of plantation	02/08/2012
Planting density (plants.ha-1)	20.833
Subscriber	07/09/2012
Start of pellet formation	30/09/2012
Date of reflection	30/10/12-09/11/12

*The beginning of pellet formation is considered to be the moment when pellets of 1 mm in diameter are detected.

2.3.2. Year 2014

Localization and initial conditions

The experience was carried out in the experimental farm of INTIA in Sartaguda (Navarra), in 2014. The Casper variety of cauliflower was used. The soil had a loam texture and its properties are shown in Table 19.

Experimental design

Transplanting was done on August 12. Irrigation was by sprinkling. As a pre-planting fertilizer, 140 kg P/ha based on 45% superphosphate and 220 kg K/ha based on 50% sulfate of potash were applied. On September 10, 28 days after transplanting (DDT), an application of nitrogen fertilizer in the form of 26% ammonium nitrosulfate was made at a variable dose depending on the available mineral nitrogen (initial Nmin + N fertilizer) (Table 20).

Table 19. Physico-chemical properties of soil in var. Casper (2014).

Dep. cm	Sand	Silt %.[1]	Clay[1] % Clay	M.O. %[2]	pH[3]	E.C.[4] dS/m	P ppm[5]	K ppm[5]	Texture[1]
0-30	53,6	34,8	11,6	0,9	8,4	0,2	40,9	262,7	Franco

| 30-60 | 53,5 | 35,1 | 11,4 | 0,7 | 8,5 | 0,2 | 25,2 | 242,9 | Franco |
| 60-90 | 54,9 | 34,5 | 10,5 | 0,4 | 8,5 | 0,2 | 1,4 | 195,8 | Franco |

1) USDA. 2) Oxidizable organic matter. 3) H_2O (1:5). 4) 25°C (1:5). 5) Mehlich III.

Four treatments were differentiated according to the available Nmin (initial Nmin + Nfertilizer) and four replicates, in a block design according to the initial Nmin. The elementary plot had an area of 81 m² and 6 crop rows (Figure 10). For the determination of mineral, nitric and ammoniacal nitrogen, present in the soil at the beginning of the crop, the depths of 0-15, 15-30 and 3060 cm were sampled. The Nmin values of the different treatments and the fertilizer doses used are shown in Table 20. A maximum rooting depth of 0.6 m was considered for the calculation of the nitrogen balance.

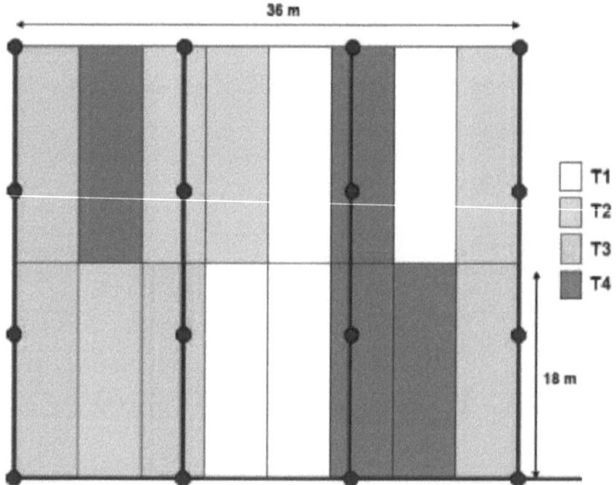

Figura 10. Diseño experimental y distribución de los tratamientos de nitrógeno mineral available in the plot of the 2014 experiment with the Casper variety. Navailable (kgN/ha): T1: 104; T2: 134; T3: 190; T4: 260. The blue Kneas and circles indicate the sprinkler network.

Experimental treatments as a function of available N in var. Casper (2014).

Treatments	T1	T2	T3	T4
Navailable (kgN/ha)	104	134	190	260
Initial Nmin (kgN/ha)	104	124	135	162
Nfertilizer (kgN/ha)	0	10	55	98

Sampling and determinations

Nmin (nitrate and ammonium) was determined at planting, 47 days after transplanting (DDT) and at harvest. For this purpose, two samples were taken in each elementary plot, mixing the soil by layers, at 0-15, 15-30 and 30-60 cm depth. At 27, 50, 68 DDT and at harvest, fresh weight, dry weight and total N were determined in leaves and pellets from five plants per elementary plot.

Harvesting took place between 99 and 110 DDT. In a subplot of 15 m^2 all the plants were collected. The total weight and the weight of the pellet were determined for each plant. All pellets of Extra or First quality and a diameter greater than 11 cm were considered as commercial (EEC, 1998).

On three dates during cultivation, leaf measurements were made with SPAD (Mod. 502, Minolta), Dualex® and Multiplex® (Force-A) equipment, at 11:00 h, on one adult leaf per plant, on ten plants per elementary plot. Five petioles of these leaves were chopped and mixed in the laboratory for sap extraction by pressing. On three dates, reflectance measurements were taken at 670, 730 and 780 nm, with the Crop-Circle™ ACS-430 (Holland Scientific). A 12-meter path was taken along the two inner lines of each elementary plot.

As in previous years, a nitrogen balance was performed. The data treatment has been similar to that of previous years. Table 21 shows a summary of some phenological and crop data.

Table 21. Casper var. trial data (2014).

Date of plantation	12/08/2014
Planting density (plants.ha-1)	20.833
Subscriber	10/09/2014
Start of pellet formation*.	01/10/2014
Collection Penodo	20/11/14-26/11/14

*The beginning of pellet formation is considered to be the moment when pellets of 1 mm in diameter are detected.

RESULTS

The results of the trials are presented by year and variety. For all the results, contained in tables and figures, the detailed values by treatment and repetition as well as their statistical significance can be found in Appendices 1, 2, 3, 4, 5, 6 and 7.

2.4. Year 2012. Var. Barcelona

Coverage, Height and Crop Biomass

Table 22 shows the results of crop height, cover and biomass at the beginning of harvest. The height and cover in the fertilized treatments were significantly higher than in the unfertilized treatment T1, with no differences between the fertilized treatments. No significant differences were found in the biomass values.

Table 22. Crop cover, height and biomass, on 10/18/2012, at the beginning of harvest.

Treatments	Height (m)		Coverage (%)		Biomass (Mg/ha)	
T1	0,49 ± 0,02	a	75 ± 5	a	3,94 ± 0,42	ns
T2	0,59 ± 0,01	b	94 ± 4	b	5,12 ± 0,31	ns
T3	0,64 ± 0,03	b	93 ± 5	b	5,33 ± 0,13	ns
T4	0,64 ± 0,03	b	87 ± 5	b	5,31 ± 0,52	ns

Different letters differ significantly in a Tukey test (p<0.05). ns: no significant differences.

Total production

The average total cauliflower production in the nitrogen fertilizer treatments was higher than 20,000 kg/ha (Table 23). Treatment T1 had a significantly lower yield.

Table 23. Total, leaf and pellet production (kg/ha) of the Barcelona variety and available nitrogen in the 2012 trial.

Treatments	Navailable	Pellas	Sheets	Total
		kg/ha		
T1	93	11.984 a	31.903 a	43.887 a
T2	189	22.142 b	46.697 b	68.839 b
T3	270	23.806 b	47.188 b	70.994 b
T4	322	22.259 b	44.954 b	67.213 b
		***	***	***

*** Significance (p<0.001) in the analysis of variance. Different letters differ significantly in a Tukey test (p<0.05).

The non-linear regression analysis of the relative total production of cauliflower as a function of available nitrogen (Ndisp = Nmin+Nfertilizer), indicates that production stabilizes for Ndisp values of 192 ± 34 kg Ndisp/ha, value of parameter *a* in the regression model (equation [3]) (Figure 11). It is understood that the treatments with a nutritional deficit state will be those that present a value of available nitrogen lower than the value at which production stabilizes. In this case, treatment T1 is the only one with available nitrogen values below the level at which production stabilizes. Treatment T2 is in the range of values above which no production increases have been obtained. Treatments T3 and T4 are well above these levels of available nitrogen.

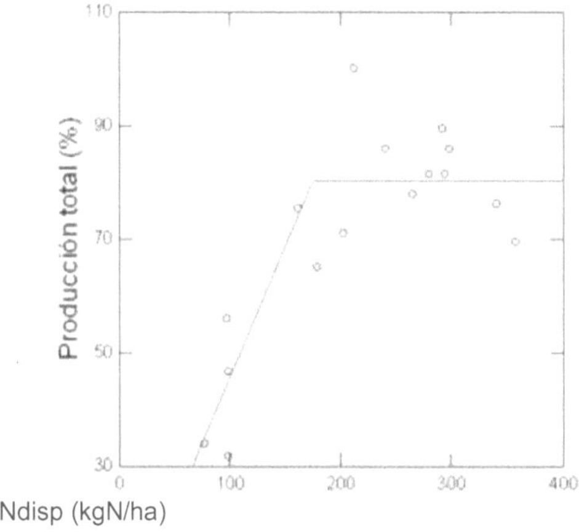

Figura 11. Relative total production of cauliflower pellets var. Barcelona, as a function of available mineral nitrogen in the soil (Nmin + N fertilizer).

Nitrogen concentration in leaves

Figure 12 shows the evolution of total nitrogen content in leaves in the different treatments. The concentration of nitrogen in cauliflower leaves decreased in all treatments as crop biomass increased. Except for the first date, the nitrogen content in the fertilized treatments was significantly higher than in the unfertilized T1 treatment.

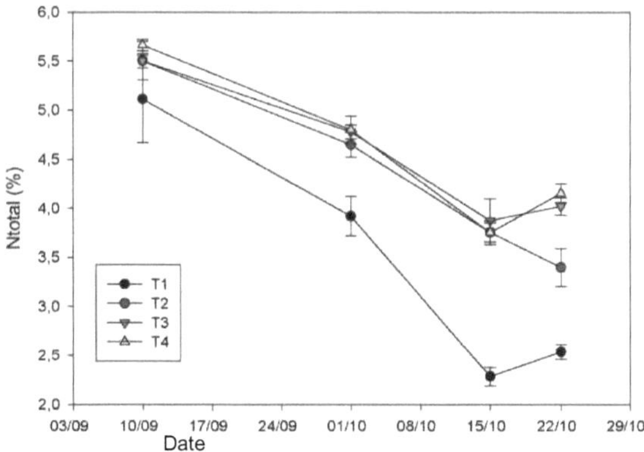

Figura 12. Total nitrogen concentration (%) of cauliflower var. Barcelona in leaves, throughout the crop in 2012. T1, T2, T3 and T4 are the treatments, with 93; 189; 270 and 322 kg/ha of Navailable.

For this trial of the Barcelona variety in 2012, according to the Greenwood (1986) model, only the T1 treatment had nitrogen concentrations below the cntic nitrogen values (Figure 13), which contributes to explain the lower production obtained for this treatment.

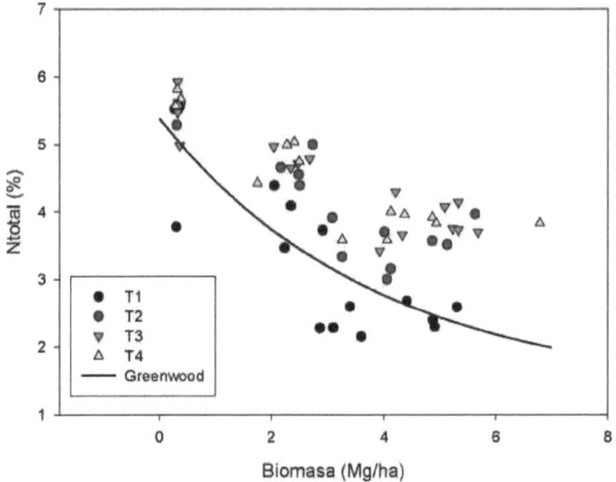

Figura 13. Total nitrogen concentration (%) of cauliflower var. Barcelona as a function of biomass (Mg/ha) in 2012. T1, T2, T3 and T4 are the treatments, with 93; 189; 270 and 322 kg available N/ha. The analytical curve of the Greenwood (1986) model is presented.

Soil nitrogen content

The initial Nmin content in the soil profile up to 0.6 m depth was between 93 and 128 kg/ha. At the end of the harvest Nmin decreased, reaching values between 6 and 73 kg/ha (Figure 14). The surface horizon appears to be almost exhausted.

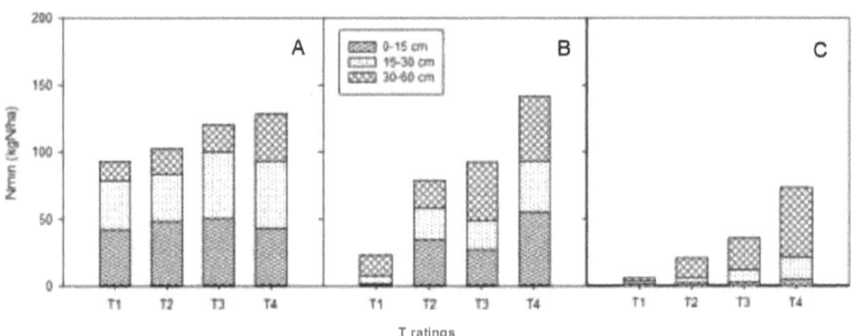

Figura 14. Mineral nitrogen (Nmin) in the soil, from 0 to 60 cm, in a crop of cauliflower var. Barcelona, in the year 2012, (A) August 12, at the time of transplanting; (B) October 1, fifteen days after mulching and (C) October 29, at the end of harvest.

N-Nitrate content in sap

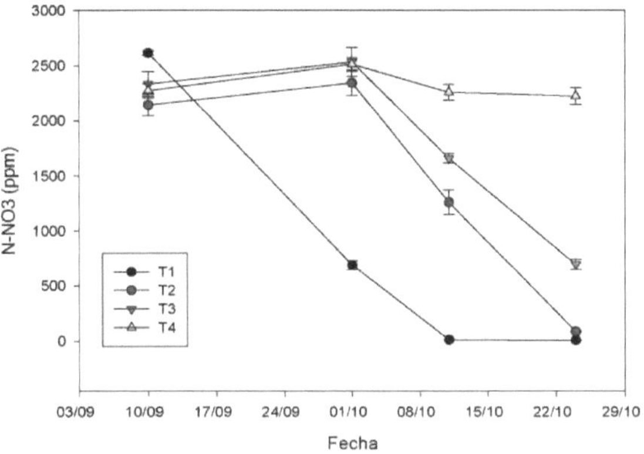

Figura 15. Concentration of N-NO3⁻ (ppm) in sap, in leaves of cauliflower var. Barcelona in the different treatments. T1, T2, T3 and T4 are the treatments with 93; 189; 270 and 322 kg available N/ha. The vertical bars indicate the standard error.

The concentration of N-NO3⁻ in sap (Figure 15) ranged between 2100 and 2600 ppm in the first measurement, two days before top dressing. This concentration increased slightly in the second determination, at the beginning of pellet formation, in treatments T2, T3 and T4, being significantly higher than in T1. Subsequently, these values decreased significantly in all treatments except T4. Concentrations in the third sampling, prior to harvest, were higher than 1000 ppm in treatments T2, T3 and T4.

SPAD sensor

In chlorophyll content, estimated with the SPAD sensor, no significant differences were found between treatments, except for the measurement on October 23 at harvest (Figure 16). No clear trends were observed over time except in treatment T1, where SPAD values gradually decreased from 63 to 54 units.

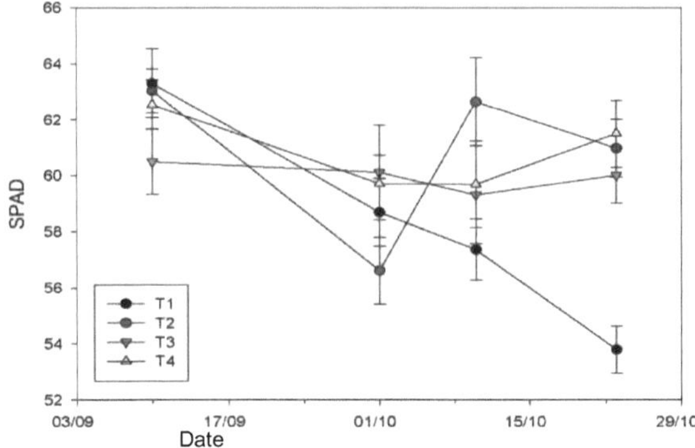

Figure 16. Chlorophyll content in leaves of cauliflower var. Barcelona, SPAD units, in the different treatments. T1, T2, T3 and T4 are the treatments, with 93; 189; 270 and 322 kg available N /ha. The vertical bars indicate the standard error.

DUALEX sensor

In the measurements made with the Dualex sensor (Figures 17 and 18), there are significant differences, from the first sampling date, for the parameters Chl and NBI between treatments compared to the least fertilized T1.

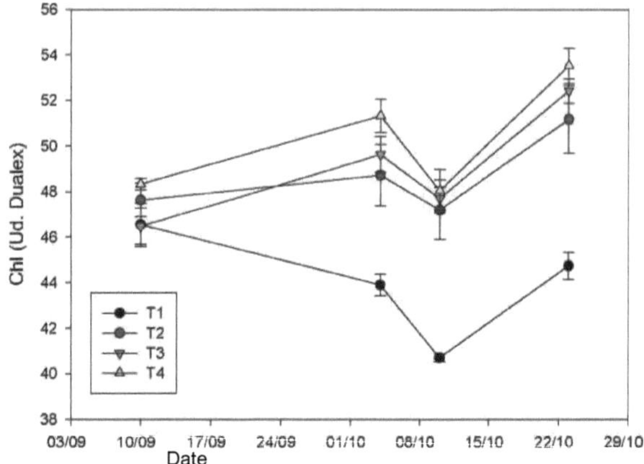

Figura 17.   Chl index of Dualex, in leaves of cauliflower var. Barcelona in the different treatments of available nitrogen. T1, T2, T3 and T4 are the treatments, with 93; 189; 270 and 322 kg available N /ha. The vertical bars indicate the standard error.

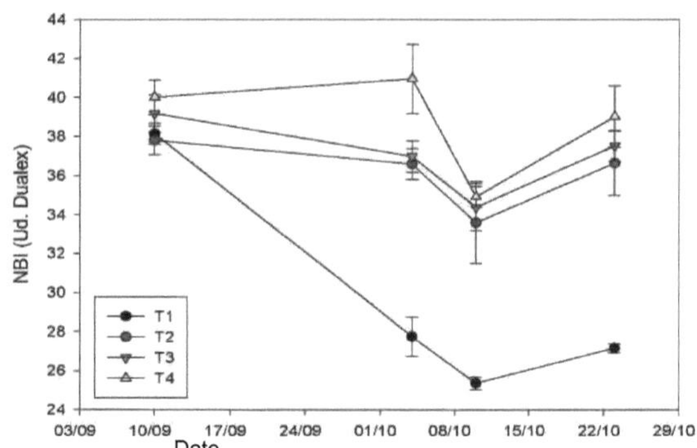

Figura 18.   Nitrogen balance index NBI of Dualex, in leaves of cauliflower var. Barcelona in the different treatments of available nitrogen. T1, T2, T3 and T4 are the treatments, with 93; 189; 270 and 322 kg available N /ha. The vertical bars indicate the standard error.

MULTIPLEX sensor

The evolution of the SFR and NBI indices of the Multiplex sensor is presented in Figures 19 and 20. At 21 days after fertilization (second sampling), the SFR index of treatment T4 is significantly higher than the rest. From this time until harvest, treatments T2, T3 and T4 show significantly higher values than T1. This treatment has shown a significantly lower NBI index than the rest as of the mulch fertilization (after the first sampling).

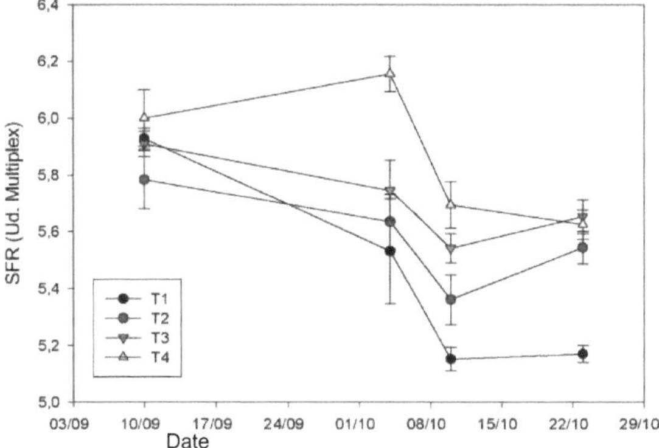

Figura 19.   fodice SFR of Multiplex, in leaves of cauliflower var. Barcelona, in the different treatments of available nitrogen. T1, T2, T3 and T4 are the treatments, with 93; 189; 270 and 322 kg available N /ha. The vertical bars indicate the standard error.

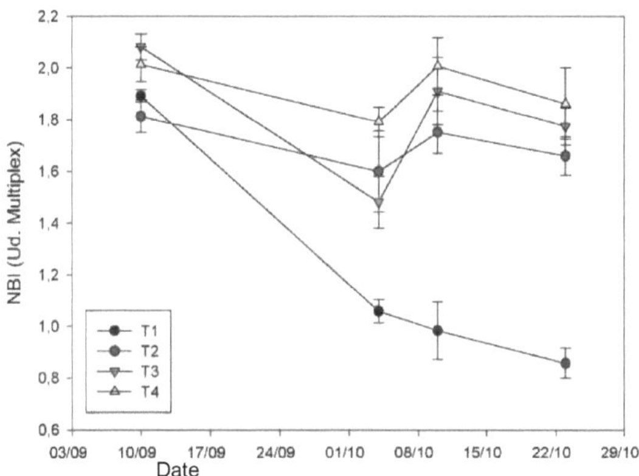

Figura 20.   fodice NBI of Multiplex, in leaves of cauliflower var. Barcelona, in the different treatments of available nitrogen. T1, T2, T3 and T4 are the treatments, with 93; 189; 270 and 322 kg available N /ha. The vertical bars indicate the standard error.

CROP CIRCLE sensor

In the trajectories carried out 49 days after planting with the Crop Circle sensor, significant differences were obtained between treatments for the NDVI index (Table 24). Treatment T1 showed a significantly lower NDRE index value than the rest.

Table 24. NDVI, NDRE indices obtained with the Crop Circle sensor in cauliflower var. Barcelona at forty-nine days after planting.

Treatments	NDVI	NDRE
T1	0,768 ± 0,012[a]	0,329 ± 0,006[a]
T2	0,788 ± 0,008[b]	0,352 ± 0,005[b]

T3	0,789 ± 0,007[b]	0,353 ± 0,004[b]
T4	0,772 ± 0,002[b]	0,350 ± 0,002[b]
	***	***

ANOVA, significance: *** (p<0.001); ns: not significant. Figures followed by different letters differ significantly. (p<0.05) in a Tukey test.

Mineralization of soil organic matter

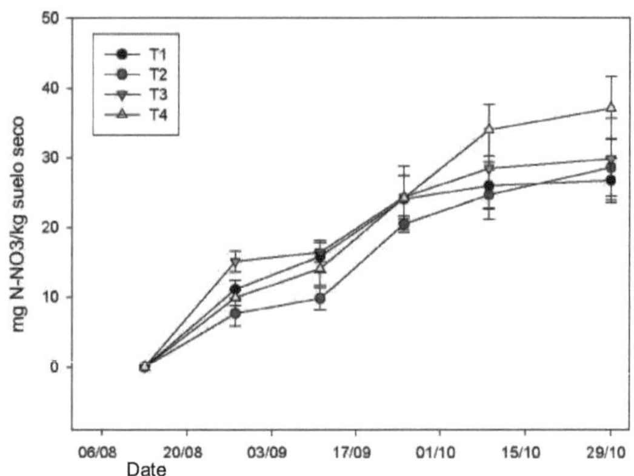

Figura 21.   Mineralization (mg N-NO3 /kg dry soil) measured in resins at 0.2 m depth, in the cauliflower trial of the Barcelona variety in 2012. Vertical bars indicate the standard error.

The mineralization rate in the 0.2 m topsoil horizon reached a mean value of 0.14 mgN/kg dry soil per day, with no significant differences between treatments (Figure 21). This value represents 41 kgN/ha for the crop penodo in the surface soil layer, extrapolated to 0.3 m depth. The higher value observed in treatment T4 could be due to an overestimation of leached N, measured by the amount of nitrate retained by the resins, caused by capillary upward movement of nitrogen.

Nitrogen balance

The results of the nitrogen balance (Table 25) indicate that the crop has been able to absorb practically all the nitrogen available in each treatment, since, taking into account the standard error in all treatments, the final balance is close to zero. In the case of treatment T1, this nitrogen was not sufficient to cover its needs, presenting a lower extraction, a lower Nmin at the end of the crop and, in turn, lower values of biomass and total nitrogen in the plant. The balance in treatment T4 may indicate higher losses due to volatilization of the nitrogen applied as fertilizer, usual in soils with pH greater than seven, in which ammonium nitrosulfate is applied, which has also caused a lower EUN in this treatment.

Nitrogen balance (kg/ha) up to 0.6 m depth.

	Nmin ini[1]	Nfert[2]	Nminer[3]	Nmin end[4]	Ncos[5]	Nlix[6]	Balance	EUN[7]
				kgN/ha				kg/kgN
T1	93±5	0	31±12	6±1[b]	121±12[b]	24±10	-27±14	89±8[a]
T2	102±12	87	34±18	21±5[b]	190±15[ab]	23±3	-10±38	100±9[a]
T3	121±11	149	31±14	36±13[ab]	256±17[a]	29±5	-20±30	86±7[a]
T4	128±16	194	70±27	73±12[a]	242±35[a]	49±7	29±24	56±3[b]
	ns	ns		**	**	ns	ns	**

1) Initial mineral N. 2) N applied as fertilizer. 3) Mineralized N from 0 to 0.3 m. 4) Final mineral N. 5) N extracted at harvest. 6) N leached from 28/09 to harvest. 7) N use efficiency: kg of commercial harvest per kg of N available. Significance: ** (p<0.01); ns: not significant. Different letters in the same column indicate significant differences (p<0.05) in a Tukey test.

4.2. Year 2013. Var. Barcelona

Coverage, Height and Crop Biomass

Table 26 shows the results of crop height, cover and biomass at the beginning of harvest. The height significantly differentiated the treatments according to the available nitrogen. Cover and biomass in the most fertilized treatments T3 and T4 were significantly higher than in the least fertilized treatments T1 and T2.

Table 26. Crop cover, height and biomass, on 09/10/2013, at the beginning of harvest.

Treatments	Height (m)		Coverage (%)		Biomass (Mg/ha)	
T1	0,43 ± 0,01	a	56 ± 5	a	2,99 ± 0,33	a
T2	0,55 ± 0,01	b	66 ± 3	ab	3,77 ± 0,35	ab
T3	0,64 ± 0,01	c	77 ± 2	bc	4,09 ± 0,18	b
T4	0,71 ± 0,01	d	82 ± 3	c	4,69 ± 0,17	b

Different letters differ significantly in a Tukey test (p<0.05). ns: no significant differences.

Total production

The average total cauliflower production in the treatment with less fertilizer (T1) exceeded 8,000 kg/ha and in the treatment with more nitrogen fertilizer (T4) it exceeded 20,000 kg/ha (Table 27). There were significant differences between treatments, depending on the available nitrogen.

Table 27. Total, leaf and pellet production (kg/ha) of the Barcelona variety and available nitrogen in the 2013 trial.

Treatments	Navailable	Pellas	Sheets	Total
		kg/ha		
T1	67	8.084 a	19.105 a	27.189 a
T2	130	13.316 b	27,114 ab	40.430 b
T3	193	17,116 bc	31,422 bc	48,538 bc
T4	260	20.748 c	36.935 c	57.683 c

		***	***	***

*** Significance (p<0.001) in the analysis of variance. Different letters differ significantly in a Tukey test (p<0.05).

The nonlinear regression analysis of relative total cauliflower production as a function of available nitrogen (Ndisp = Nmin+Nfertilizer) indicates that production stabilizes for values of 178 ± 37 kg Ndisp/ha (Figure 22), value of parameter *a* in the regression model [equation 3]. In this case, treatments T1 and T2 present values of available nitrogen lower than the level at which production stabilizes. And treatments T3 and T4 present values higher than the level at which production stabilizes.

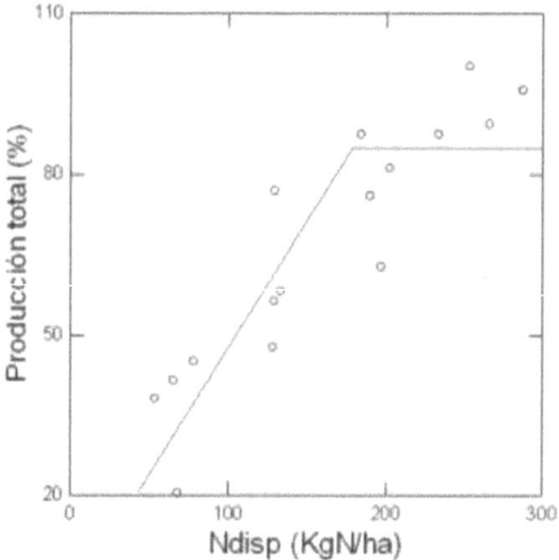

Figura 22.   Relative total production of cauliflower pellets var. Barcelona, as a function of available nitrogen in the soil (Nmin + N fertilizer).

Nitrogen concentration in leaves

The nitrogen content in cauliflower leaves (Figure 23) maintained values similar to the initial values until the first days of October, in the third sampling, possibly due to the decrease in biomass that occurred after a hail event. Subsequently, it decreased in all treatments as crop biomass increased, differentiating at harvest according to the available nitrogen (Figure 24). According to the Greenwood (1986) model, and for biomass values higher than 1 Mg/ha, treatment T1 had nitrogen concentrations below the cntic nitrogen values and this would justify the lower production values obtained. Treatment T2, which also had a lower production, had total nitrogen values above the critical curve, but close to the deficit zone that differentiates the model.

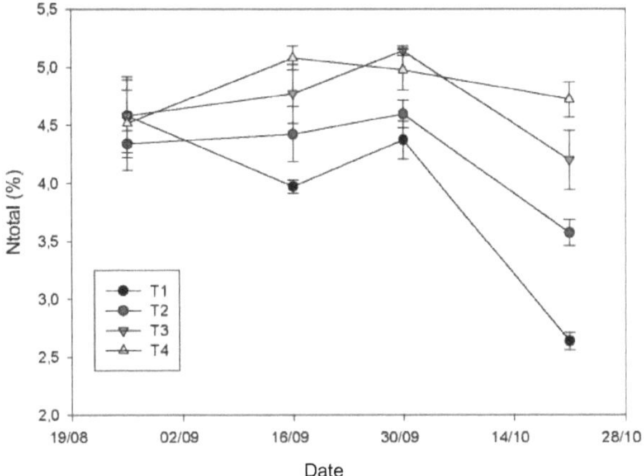

Figura 23.   Total nitrogen concentration (%) of cauliflower var. Barcelona in leaves, throughout the crop in 2013. T1, T2, T3 and T4 are the treatments, with 67; 130; 193 and 260 kg available N /ha.

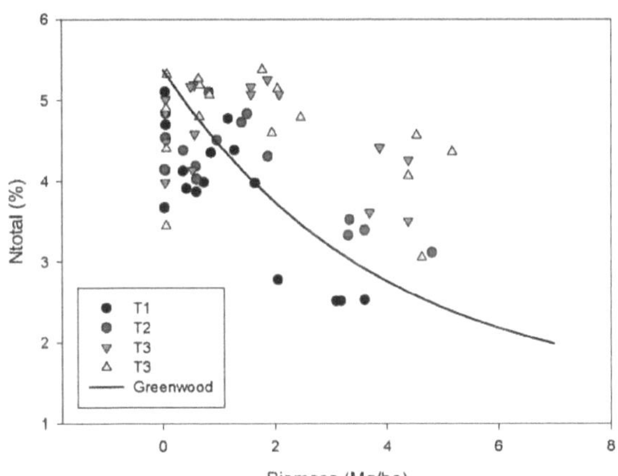

Figura 24.   Total nitrogen concentration (%) in cauliflower var. Barcelona as a function of biomass (Mg/ha) in 2013. T1, T2, T3 and T4 are the treatments, with 67; 130; 193 and 260 kg available N/ha. The analytical curve of the Greenwood 1986 model is presented.

Soil nitrogen content

The initial Nmin content (August 5) in the soil profile up to 0.6 m depth was between 67 and 130 kg/ha (Figure 25). At the beginning of pellet formation, 28 days after top dressing fertilization, the Nmin content decreased by 46% in T1, 48% in T2, 18% in T4 and increased by 20% in T3 (Figure 25). At the end of the harvest, Nmin decreased in all treatments, below 50 kg/ha. The surface horizon appears to be almost depleted in all treatments.

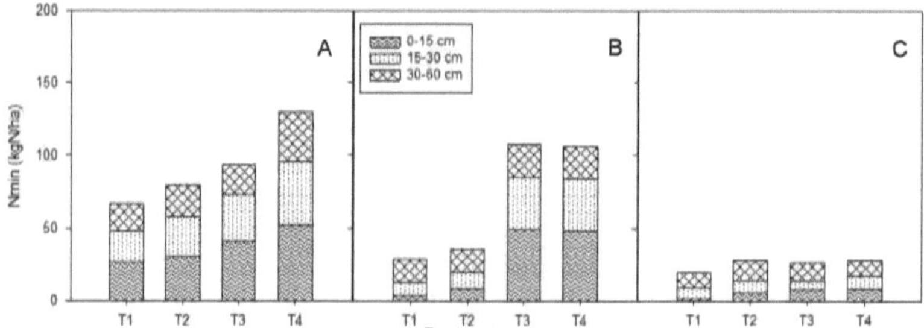

Figure 25. Mineral nitrogen (Nmin) in the soil (kgN/ha) from 0 to 60 cm depth at the var. Barcelona, in 2013, (A) August 5, at the time of transplantation, (B) August 30, at the time of transplantation, (C) August 30, at the time of transplantation, (D) August 30, at the time of transplantation.

September, at the beginning of pellet formation and (C) October 28, at the end of harvest.

N-Nitrate content in sap

In the first sampling, three days before the top dressing, the concentration of $N\text{-}NO_3^-$ in sap ranged between 1,200 and 1,300 ppm (Figure 26), with no significant differences among treatments. In the second sampling, this concentration decreased, possibly due to the hail event that affected the plants, and in the third sampling, at the beginning of pellet formation, it recovered, with the fertilized treatments being significantly higher than the unfertilized ones.

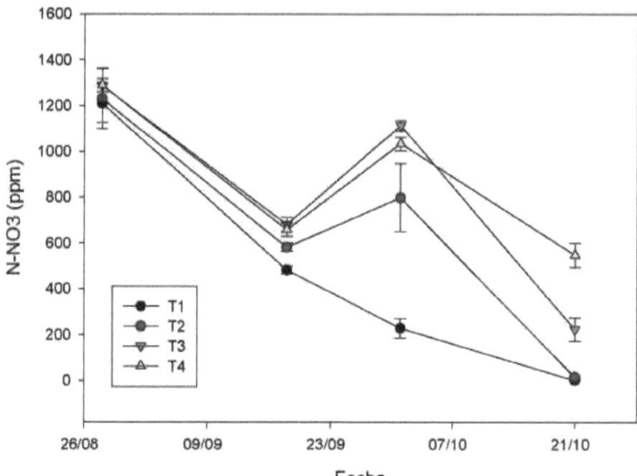

Figure 26. Concentration of $N\text{-}NO_3^-$ (ppm) in sap, in leaves of cauliflower var. Barcelona in the different treatments. T1, T2, T3 and T4 are the treatments with 67; 130; 193 and 260 kg N available/ha. The vertical bars indicate the standard error.

SPAD sensor

In the measurements made with the SPAD sensor (Figure 27), no significant differences were found before the top dressing fertilization, which was carried out two days before the second sampling. The evolution of its values was similar to that of nitrate in sap. In the fourth sampling, at the beginning of pellicle formation, the unfertilized T1 treatment differed significantly from the more fertilized T4. In the last sampling, at harvest, significant differences were found between the least fertilized treatments, T1 and T2, and the most fertilized treatments, T3 and T4.

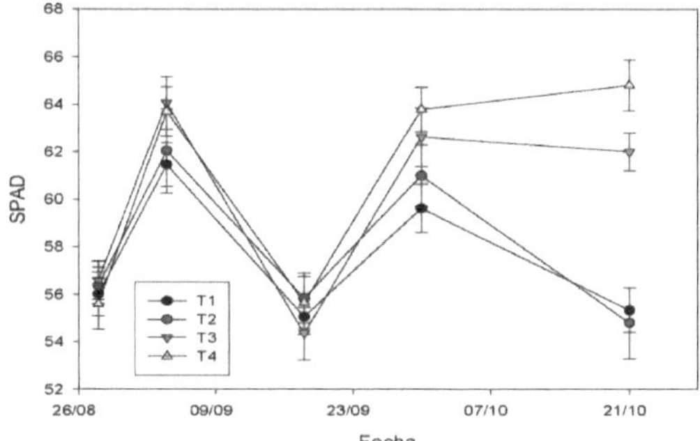

Figure 27. Chlorophyll content in leaves of cauliflower var. Barcelona, SPAD units, in the different treatments. T1, T2, T3 and T4 are the treatments, with 67; 130; 193 and 260 kg available N /ha. The vertical bars indicate the standard error.

DUALEX sensor

In the measurements made with the DUALEX sensor, there were significant differences for the Chl and NBI indices between treatments (Figures 28 and 29) from the cover fertilization, which was made two days before the second sampling, differentiating the T4 treatment from the rest. In the fourth sampling, at the beginning of pellet formation, and in the fifth, at harvest, significant differences were found between the least fertilized and most fertilized treatments.

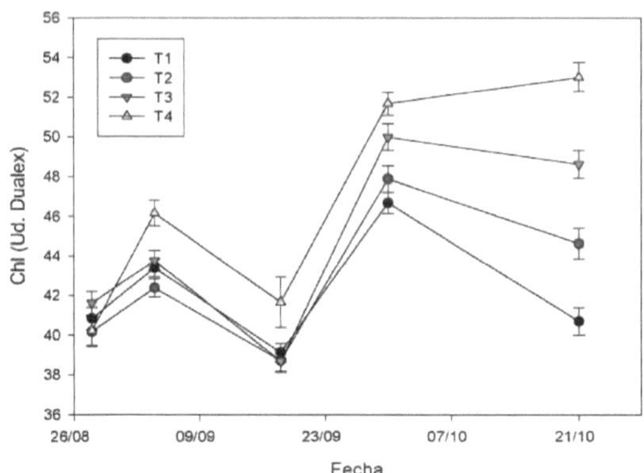

Figura 28.   Chl index of Dualex, in leaves of cauliflower var. Barcelona, in the different treatments of available nitrogen. T1, T2, T3 and T4 are the treatments with 67; 130; 193 and 260 kg available N/ha. The vertical bars indicate the standard error.

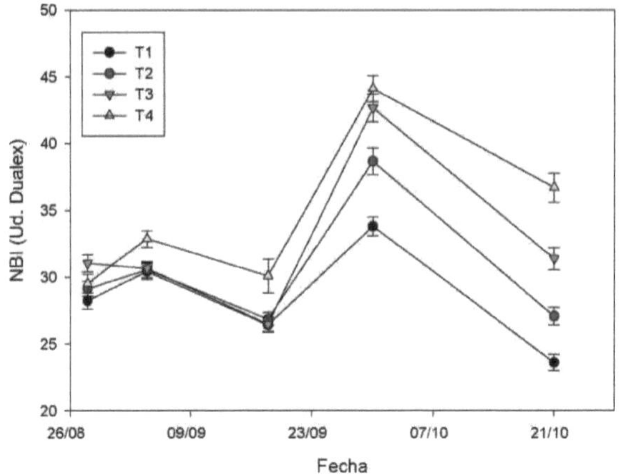

Figura 29. Nitrogen balance index NBI of Dualex, in leaves of cauliflower var. Barcelona, in the different treatments of available nitrogen. T1, T2, T3 and T4 are the treatments with 67; 130; 193 and 260 kg available N/ha. The vertical bars indicate the standard error.

MULTIPLEX sensor

The evolution of the SFR and NBI indices of the MULTIPLEX sensor is shown in Figures 30 and 31. There are significant differences in both indices. In the fourth sampling, at 26 days after top dressing, the SFR index of the most fertilized treatment, T4, differs significantly from the rest. At harvest, the most fertilized treatments, T3 and T4, differed from the unfertilized treatment T1. For the NBI index, in the fourth sampling, at 26 days after mulch fertilization, treatment T1 is significantly lower than the rest. At harvest, the most fertilized treatment, T4, is significantly different from the rest.

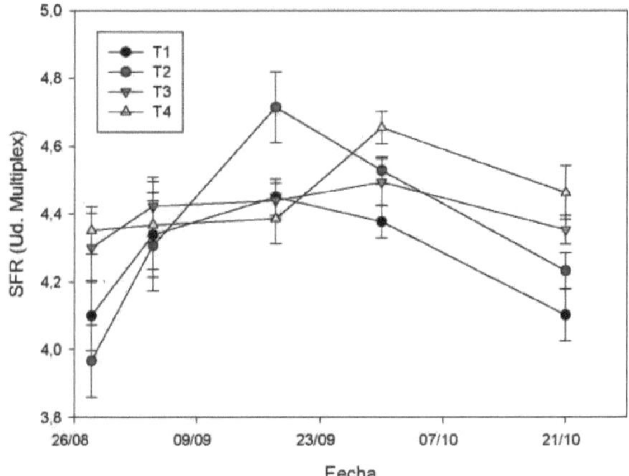

Figura 30. Multiplex SFR index, in leaves of cauliflower var. Barcelona, in the different treatments of available nitrogen. T1, T2, T3 and T4 are the treatments with 67; 130; 193 and 260 kg available N/ha.

The vertical bars indicate the standard error.

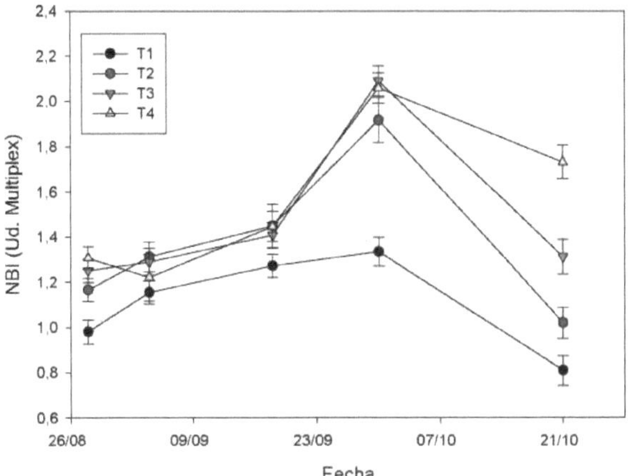

Figura 31.   Multiplex nitrogen balance index NBI, in leaves of cauliflower var. Barcelona, in the different treatments of available nitrogen. T1, T2, T3 and T4 are the treatments with 67; 130; 193 and 260 kg available N/ha. The vertical bars indicate the standard error.

CROP CIRCLE sensor

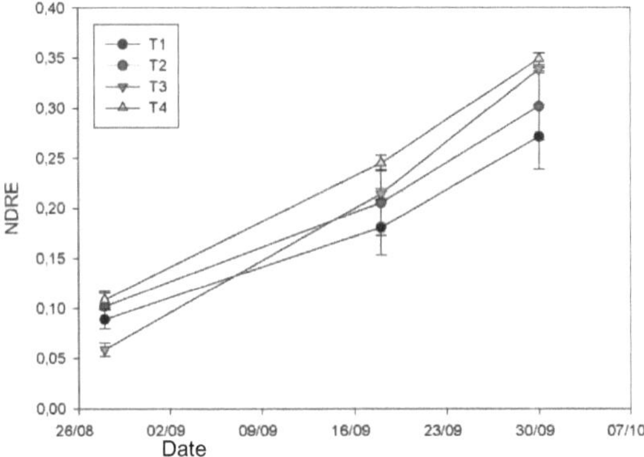

Figura 32.   fodice NDRE of Crop Circle, in leaves of cauliflower var. Barcelona, in the different treatments of available nitrogen. T1, T2, T3 and T4 are the treatments, with 67; 130; 193 and 260 kg available N/ha. The vertical bars indicate the standard error.

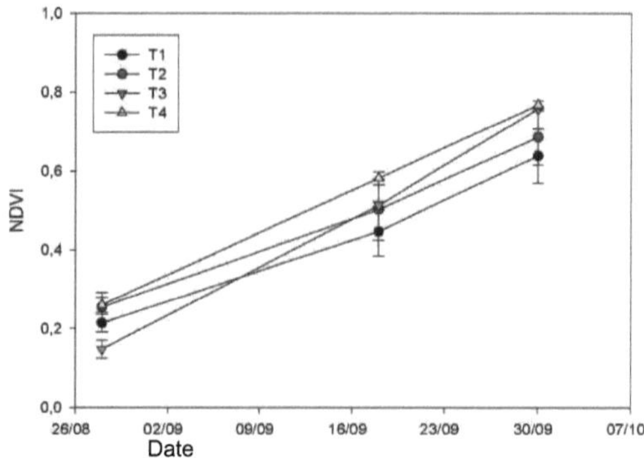

Figura 33. NDVI fodice of Crop Circle, in leaves of cauliflower var. Barcelona, in the different treatments of available nitrogen. T1, T2, T3 and T4 are the treatments, with 67; 130; 193 and 260 kg available N/ha. The vertical bars indicate the standard error.

In the measurements made with the Crop Circle sensor, significant differences were obtained for the NDRE and NDVI indices, among the treatments, before the mulch fertilization that was carried out after the second sampling (Figures 32 and 33). The value of the indices is higher in the T4 treatment and decreases in the rest of the treatments, depending on the available nitrogen.

Mineralization of soil organic matter

The mineralization rate in the 0.2 m topsoil horizon reached a mean value of 0.18 mgN/kg dry soil per day (Figure 34), with no significant differences between treatments T1 and T4. This value represents 41 kgN/ha for the crop penodo, extrapolated to the topsoil layer up to 0.3 m.

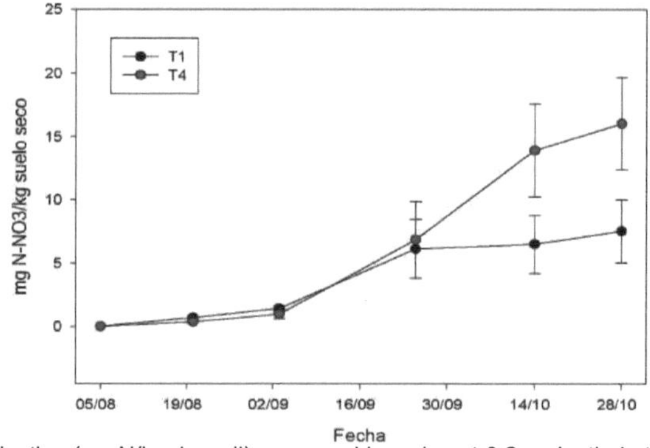

Figure 34. Mineralization (mg N/kg dry soil) measured in resins at 0.2 m depth, in the 2013 Barcelona variety cauliflower trial, in treatment T1 (67 kg available N/ha) and T4 (260 kg available N/ha). The vertical bars indicate the standard error.

Nitrogen balance

In the case of treatments T1 and T2, the available nitrogen was not sufficient to cover their needs, presenting a lower nitrogen extraction by the crop, due to a lower biomass and a lower concentration of total nitrogen. The balance in treatment T4 may indicate losses due to volatilization of nitrogen applied as fertilizer, which is common in soils with pH greater than 7, where ammonium nitrosulfate is applied (Table 28).

Table 28. Nitrogen balance (kg N/ha) up to 0.6 m depth.

	Nmin ini[1]	Nfert[2]	Nminer[3]	Nmin end[4]	Ncos[5]	Nlix[6]	Balance	EUN[7]
				kgN/ha				kg/kgN
T1	67±5[a]	0		20±8	77±7[a]	12±2	-1±3[a]	76±11
T2	80±1[a]	50	41±9	28±8	125±9[b]	16±4	1±9[a]	78±8
T3	93±4[a]	100		27±11	161±12[b]	14±1	33±20[ab]	74±6
T4	130±11[b]	130		28±6	201±10[c]	16±5	55±9[b]	69±3
	***			ns	***	ns	*	ns

Year 2014. Var. Barcelona

Coverage, Height and Crop Biomass

Table 29 shows the results of crop height, cover and biomass at the beginning of harvest. The height significantly differentiated the treatments according to the amount of nitrogen available. The cover in the most fertilized treatment was significantly higher than in the unfertilized treatment T1. No significant differences were found in biomass values.

Table 29. Crop cover, height and biomass, on 10/22/14, at the beginning of harvest.

Treatments	Height (m)		Coverage (%)		Biomass (Mg/ha)	
T1	0,51 ± 0,01	a	67 ± 2	a	3,70 ± 0,24	ns
T2	0,56 ± 0,01	b	75 ± 3	ab	3,96 ± 0,21	ns
T3	0,62 ± 0,01	c	77 ± 3	ab	4,11 ± 0,21	ns
T4	0,71 ± 0,01	d	79 ± 2	b	4,49 ± 0,23	ns

Different letters differ significantly in a Tukey test (p<0.05). ns: no significant differences.

Total production

The average total production of cauliflowers in the T3 and T4 treatments, plus fertilizers, was of the order of 16,000 kg/ha (Table 30). There were significant differences between treatments, depending on the available nitrogen.

---

1) Initial mineral N. 2) N applied as fertilizer. 3) Mineralized N from 0 to 0.3 m. 4) Final mineral N. 5) N extracted at harvest. 6) N leached from 28/09 to harvest. 7) N use efficiency: kg of commercial harvest per kg of N available. Significance: *** (p<0.001); * (p<0.05); ns: not significant. Different letters in the same column indicate significant differences (p<0.05) in a Tukey test.

Total, leaf and pellet production (kg/ha) of the Barcelona variety and available nitrogen in the 2014 trial.

Treatments	Navailable	Pellas	Sheets	Total
	kg/ha			
T1	72	7.328 a	24.201 a	31.529 a
T2	130	11,068 ab	30.553 b	41,621 ab
T3	189	14,517 bc	34,952 bc	49,470 bc
T4	260	17.318 c	38.965 c	56.283 c
		***	***	***

*** Significance (p<0.001) in the analysis of variance. Different letters differ significantly in a Tukey test (p<0.05).

Non-linear regression analysis of relative total cauliflower production as a function of available nitrogen (Ndisp = Nmin+Nfertilizer) indicates that production stabilizes for Ndisp values of 179 ± 41 kg Ndisp/ha (Figure 35). In this case, treatments T1 and T2 are below this level of available nitrogen, and treatments T3 and T4 are above it.

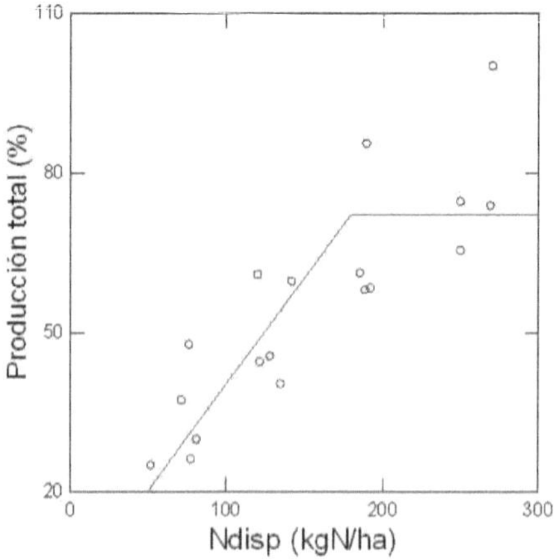

Figura 35. Relative total production of cauliflower pellets var. Barcelona, as a function of available nitrogen in the soil (Nmin + N fertilizer).

Nitrogen concentration in leaves

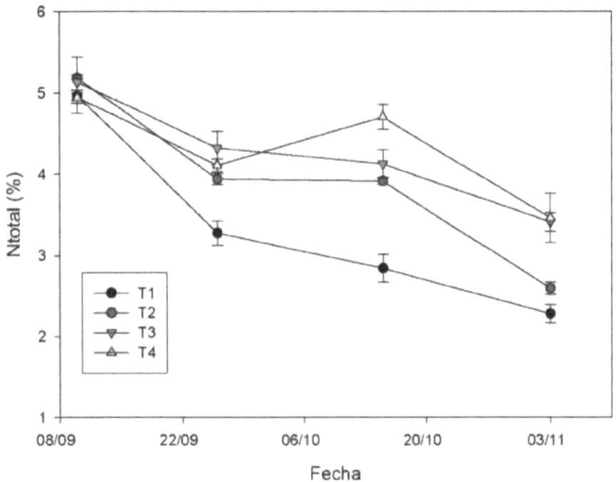

Figura 36. Total nitrogen concentration (%) of cauliflower var. Barcelona in leaves, throughout the crop in 2014. T1, T2, T3 and T4 are the treatments, with 72; 130; 189 and 260 kg available N /ha.

The nitrogen content in cauliflower leaves in the first sampling, the day before top dressing fertilization, was around 5 % in all treatments (Figure 36). In the second sampling, after top dressing fertilization, nitrogen concentrations decreased until harvesting, with a greater decrease in the less fertilized treatments T1 and T2, with significant differences between treatments. Treatments T3 and T4 reached values of 3.5% at the end of the crop and treatments T1 and T2 reached values of 2.5% at that time. According to the Greenwood (1986) model, it was treatments T1 and T2 that had nitrogen concentrations below the cntic nitrogen values (Figure 37).

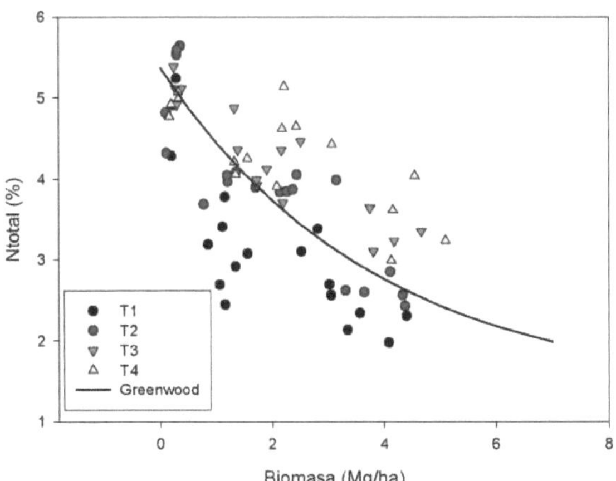

Figure 37. Total nitrogen concentration (%) of cauliflower var. Barcelona as a function of biomass (Mg/ha) in 2014. T1, T2, T3 and T4 are the treatments, with 72; 130; 189 and 260 kg available N /ha. The critical curve of the Greenwood 1986 model is presented.

Soil nitrogen content

The initial Nmin content (August 5) in the soil profile up to 0.6 m depth was between 70 and 120 kg/ha (Figure 38). Eighteen days after top dressing fertilization, the Nmin content decreased in all treatments except in T4, where it increased. At the end of harvesting, Nmin decreased in all treatments, below 25 kg/ha. The surface horizon up to 0.15 m appears almost depleted.

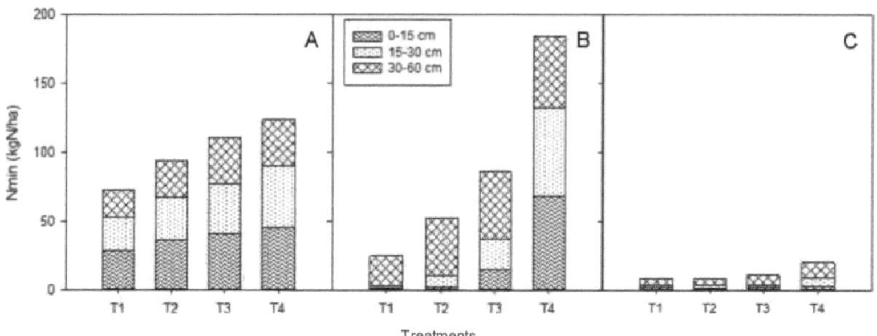

Figura 38. Mineral nitrogen (Nmin) in the soil, from 0 to 60 cm, in the var. Barcelona, in the year 2014, (A) August 5, at the time of transplanting, (B) September 29, eighteen days after fertilization and (C) November 10, at the end of harvest.

N-Nitrate content in sap

In the first sampling, the day before top dressing, the concentration of N-NO3⁻ in sap ranged between 1,600 and 1,800 ppm (Figure 39), with no significant differences between treatments. This concentration decreased to values below 800 ppm in the second sampling, eighteen days after top dressing, with significant differences between treatments.

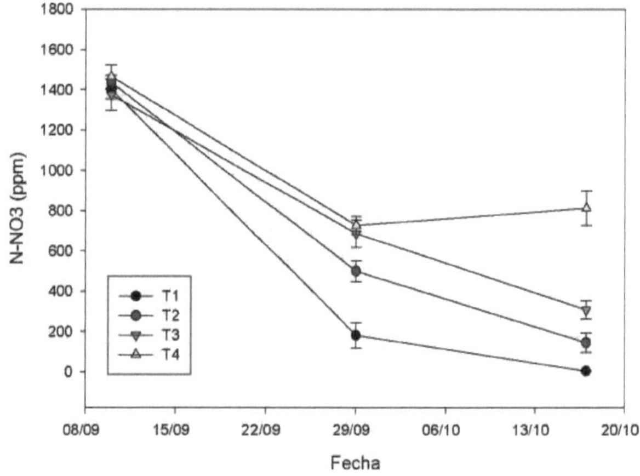

Figura 39. Concentration of N-NO3⁻ (ppm) in sap, in leaves of cauliflower var. Barcelona in the different treatments. T1, T2, T3 and T4 are the treatments with 72; 130; 189 and 260 kg available N/ha. The vertical bars indicate the standard error.

Nitrate concentration continued to decrease to values below 400 ppm in the third sampling, at the beginning of the reflection, in all treatments except the most fertilized T4.

SPAD sensor

In the measurements made with the SPAD sensor (Figure 40), no significant differences were found in the first sampling, before the mulch fertilization, nor in the second sampling, eighteen days later.

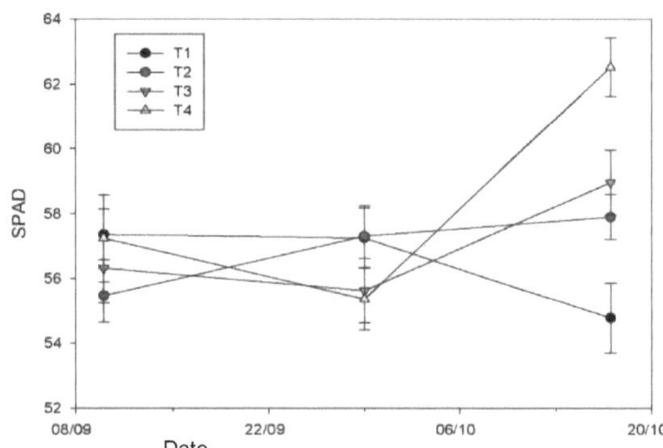

Figure 40. Chlorophyll content in leaves of cauliflower var. Barcelona, SPAD units, in the different treatments. T1, T2, T3 and T4 are the treatments, with 72; 130; 189 and 260 kg available N /ha. The vertical bars indicate the standard error.

In the third sampling, prior to harvest, significant differences were found between treatments, with the most fertilized T4 showing the highest SPAD values and the unfertilized T1 the lowest.

DUALEX sensor

In the measurements made with the DUALEX sensor (Figures 41 and 42), no significant differences were found for the Chl and NBI indices in the first sampling, before top dressing fertilization, nor in the second, eighteen days later. In the third sampling, before harvest, significant differences were found between treatments, with the most fertilized T4 showing the highest values for these indices and the unfertilized T1 the lowest.

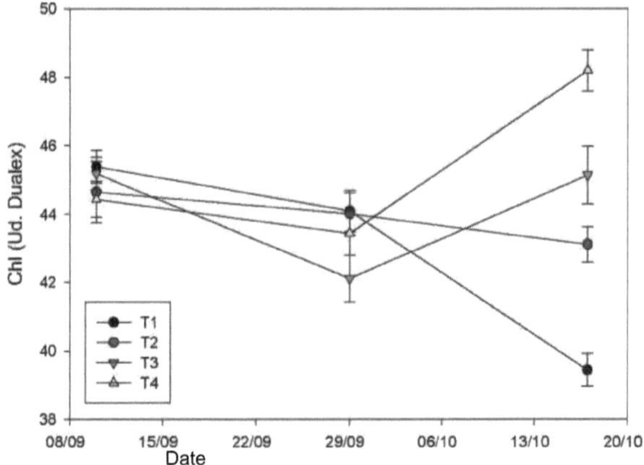

Figura 41. Íodice Chl of Dualex, in cauliflower cv. Barcelona, in the different treatments of available nitrogen. T1, T2, T3 and T4 are the treatments, with 72; 130; 189 and 260 kg available N /ha. The vertical bars indicate the standard error.

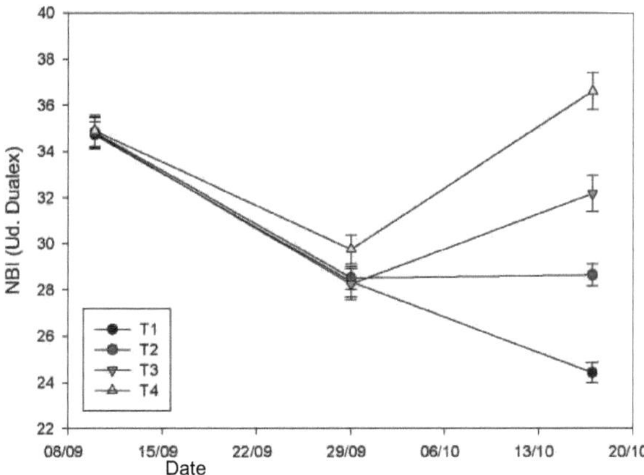

Figura 42. Íodice NBI of Dualex, on cauliflower cv. Barcelona, in the different treatments of available nitrogen. T1, T2, T3 and T4 are the treatments, with 72; 130; 189 and 260 kg available N /ha. The vertical bars indicate the standard error.

MULTIPLEX sensor

The evolution of the SFR and NBI indices of the MULTIPLEX sensor is shown in Figures 43 and 44. There are significant differences with the NBI index at all dates and with the SFR index from the top dressing fertilization, after the first sampling. The evolution of the values of these indices has been similar in all treatments, with the lowest values being obtained in the least fertilized treatments T1 and T2.

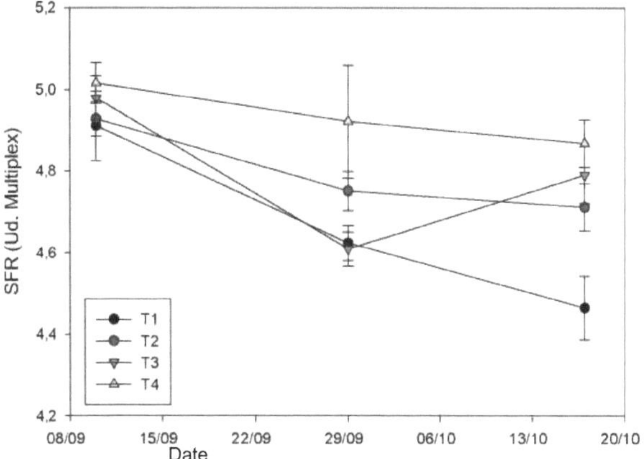

Figura 43. SFR index of Multiplex, in cauliflower cv. Barcelona, in the different treatments of available nitrogen. T1, T2, T3 and T4 are the treatments, with 72; 130; 189 and 260 kg N available /ha. The vertical bars indicate the standard error.

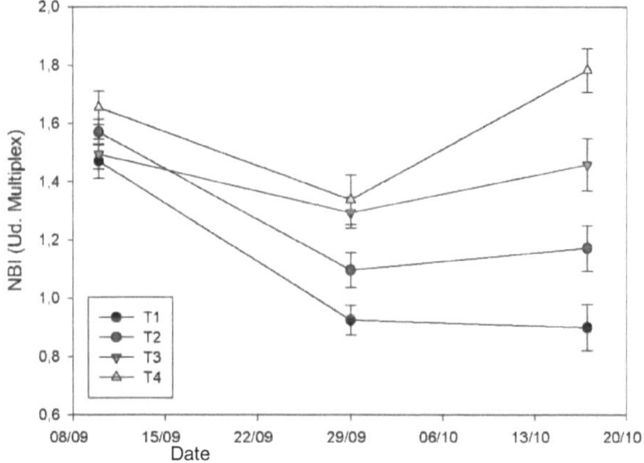

Figura 44. Multiplex NBI index in cauliflower cv. Barcelona, in the different treatments of available nitrogen. T1, T2, T3 and T4 are the treatments with 72; 130; 189 and 260 kg N available/ha. The vertical bars indicate the standard error.

CROP CIRCLE sensor

In the trajectories performed with the Crop Circle sensor, significant differences were obtained for the NDRE and NDVI indices between treatments, for each sampling date (Figures 45 and 46). The highest values of these indices were obtained for the most fertilized treatments and the lowest for the least fertilized treatments.

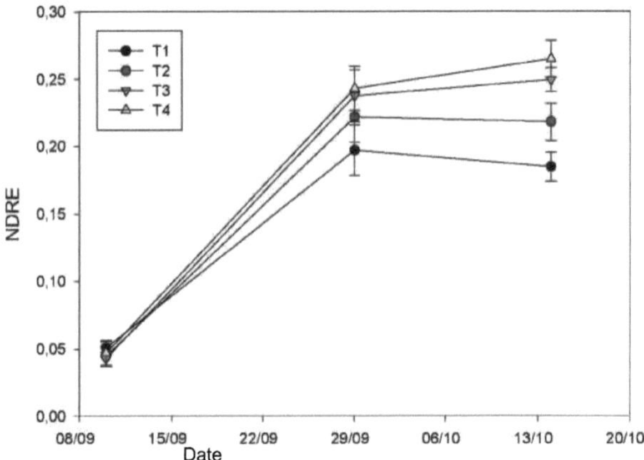

Figura 45. NDRE index of Crop Circle, in cauliflower cv. Barcelona in the different treatments of available nitrogen. T1, T2, T3 and T4 are the treatments, with 72; 130; 189 and 260 kg available N /ha. The vertical bars indicate the standard error.

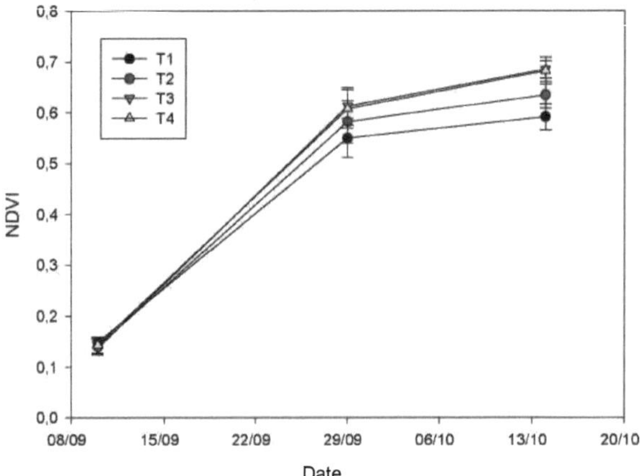

Figura 46. Crop Circle NDVI index, in cauliflower cv. Barcelona in the different treatments of available nitrogen. T1, T2, T3 and T4 are the treatments, with 72; 130; 189 and 260 kg available N /ha. The vertical bars indicate the standard error.

Mineralization of soil organic matter

The rate of mineralization in the 0.2 m topsoil horizon reached a mean value of 0.22 mgN/kg dry soil per day (Figure 47), with no significant differences between treatments. This value represents 57 kgN/ha for the crop penodo, extrapolated to the topsoil up to 0.3 m.

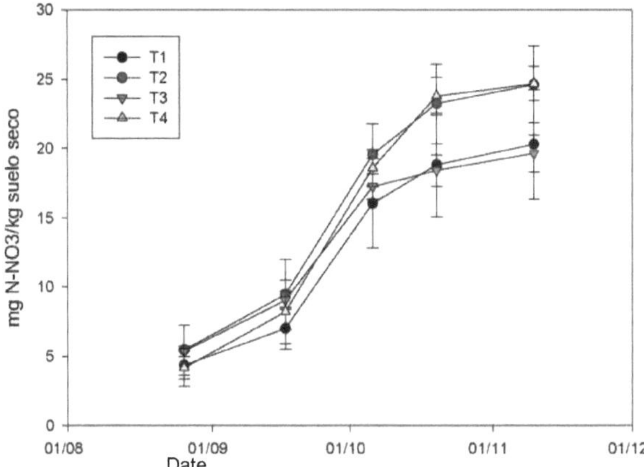

Mineralization (mg N/kg dry soil) measured in resins at 0.2 m depth, in the cauliflower trial of the Barcelona variety in 2014, in treatments T1, T2, T3 and T4 with 72; 130; 189 and 260 kg available N /ha. The vertical bars indicate the standard error.

Nitrogen balance

From the terms of the balance it can be deduced that there is nitrogen in excess, especially in the most fertilized treatments, T3 and T4. This could indicate losses due to volatilization of the nitrogen applied as fertilizer, which is common in soils with pH greater than seven, where ammonium nitrosulfate is applied. In the case of treatments T1 and T2, the available nitrogen was not sufficient to cover their needs, presenting a lower nitrogen extraction by the crop. No significant differences were found in the efficiency of nitrogen use (Table 31).

Table 31. Nitrogen balance (kg/ha) up to 0.6 m depth

	Nmin ini	Nfert	Nminer[ll23]	Nmin fin	Ncos	Nlix[4] [56]	Balance	EUN[7]
			kgN/ha					kg/kgN
T1	72±5a	0	57±3	8±1a	83±6a	17±2a	21±8a	57±7
T2	94±4b	36	57±3	8±2a	104±6a	24±2ab	50±5ab	59±5
T3	110±1bc	79	57±3	11±2a	136±8b	25±2b	75±10b	59±6
T4	124±6c	136	57±3	20±4b	155±13b	24±1ab	119±12c	55±5
	***			**	***	*	***	ns

4.4. Year 2013. Var. Typical

Coverage, Height and Crop Biomass

Table 32 shows the results and their significant differences, by treatments, for height, cover and biomass of the crop at the beginning of harvest. The height significantly differentiated the treatments according to the amount of nitrogen available. Cover in the most fertilized treatment was significantly higher than in the unfertilized treatment T1. No significant differences were found in biomass values.

Table 32. Crop cover, height and biomass, on 01/21/2014, at the beginning of harvest.

Treatments	Height (m)		Coverage (%)		Biomass (Mg/ha)	
T1	0,56 ± 0,01	a	81 ± 4	a	5,93 ± 0,50	ns
T2	0,62 ± 0,01	b	89 ± 4	ab	6,43 ± 0,39	ns
T3	0,68 ± 0,01	c	92 ± 1	ab	6,76 ± 0,33	ns
T4	0,70 ± 0,01	c	93 ± 1	b	7,43 ± 0,21	ns

Different letters differ significantly in a Tukey test (p<0.05). ns: no significant differences.

Total production

The average total cauliflower production in treatments T3 and T4, plus fertilizers, was of the order of 20,000 kg/ha (Table 33). There were significant differences between treatments in terms of available nitrogen.

Table 33. Total, leaf and pellet yield (kg/ha) of Typical and available nitrogen in the 2013 trial.

Treatments	Navailable	Pellas	Sheets	Total
		kg/ha		
T1	84	10.035 a	32.066 a	42.101 a
T2	130	14,113 ab	37,135 ab	51.248 a
T3	190	18,910 bc	45,635 ab	64.545 a
T4	260	22.392 c	50.811 b	73.203 b
		**	**	**

** Significance (p<0.01) in the analysis of variance. Different letters differ significantly in a Tukey test (p<0.05).

The non-linear regression analysis of relative total cauliflower production as a function of available nitrogen (Ndisp = Nmin+Nfertilizer) indicates that production stabilizes for values of 189 ± 45 kg Ndisp/ha (Figure 48). In this case, treatments T1 and T2 are below this level of available nitrogen, while treatments T3 and T4 are above this level of available nitrogen.

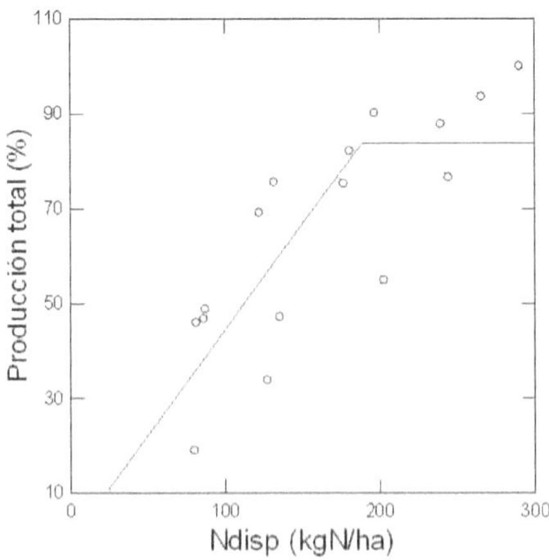

Figura 48. Relative total production of cauliflower pellets var. Typical, as a function of available nitrogen in the soil (Nmin + N fertilizer).

Nitrogen concentration in leaves

The nitrogen content in cauliflower leaves was maintained at around 4.5% in treatments T3 and T4 until the third sampling, 12 days after the second top dressing fertilization (Figure 49). In treatments T1 and T2 this concentration was maintained until the second sampling, 14 days after the first top dressing fertilization. Only in the third sampling were significant differences found between treatments, differentiating the unfertilized treatment, T1, from the rest. Subsequently, nitrogen concentration decreased gradually to 3% in treatments T3 and T4 and 2% in treatments T1 and T2 in the fifth sampling, at the beginning of harvesting. According to the Greenwood (1986) model, treatments T1 and T2 had nitrogen concentrations below the cntic nitrogen values (Figure 50).

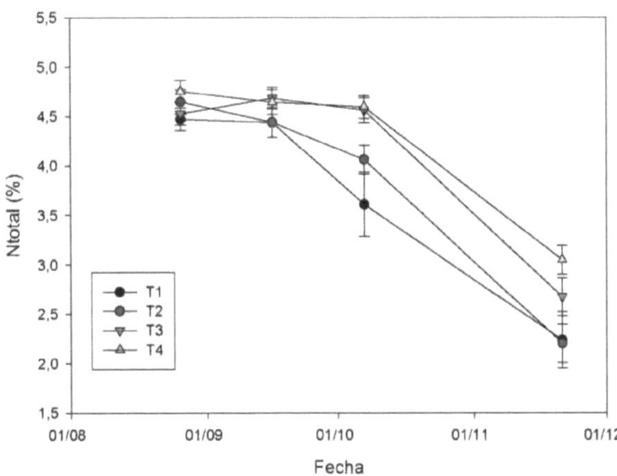

Figura 49. Total nitrogen concentration (%) of cauliflower var. Typical in leaves, throughout the crop in 2013. T1, T2, T3 and T4 are the treatments, with 84; 130; 190 and 260 kg available N /ha.

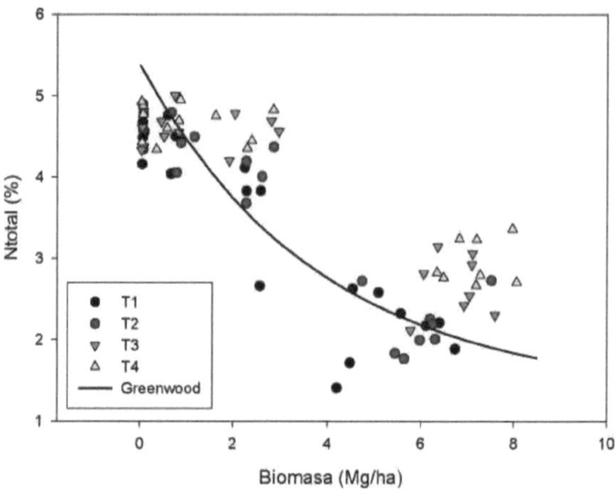

Figura 50. Total nitrogen concentration (%) in cauliflower var. Typical as a function of biomass (Mg/ha) in 2013. T1, T2, T3 and T4 are the treatments, with 84; 130; 190 and 260 kg available N /ha. The analytical curve of the Greenwood 1986 model is presented.

Soil nitrogen content

The initial Nmin content (August 6) in the soil profile up to 0.6 m depth ranged from 80 to 180 kg/ha (Figure 51). At fifteen days   The Nmin content decreased to 60 kg/ha in treatment T1, was maintained in treatments T2 and T3, and increased to over 200 kg/ha in T4. At the end of the harvest, Nmin decreased in all treatments, below 20 kg/ha. The surface horizon up to 0.15 m appears almost depleted.

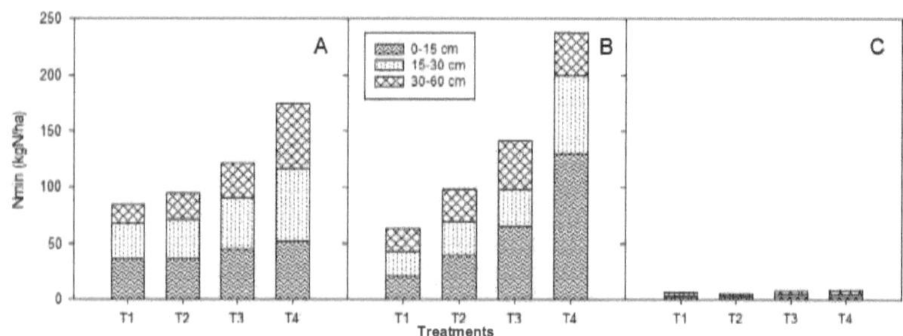

Figure 51. Mineral nitrogen (Nmin) in the soil, from 0 to 60 cm, in the var. Typical, in the anus. 2013, (A) August 6, at the time of transplantation, (B)September 16, fifteen d^as. after the first fertilization and (C) March 10, at the end of harvest.

N-Nitrate content in sap

In the first sampling, four days before top dressing, the concentration of N-NO3⁻ in sap ranged between 1,200 and 1,400 ppm, with no significant differences between treatments (Figure 52).

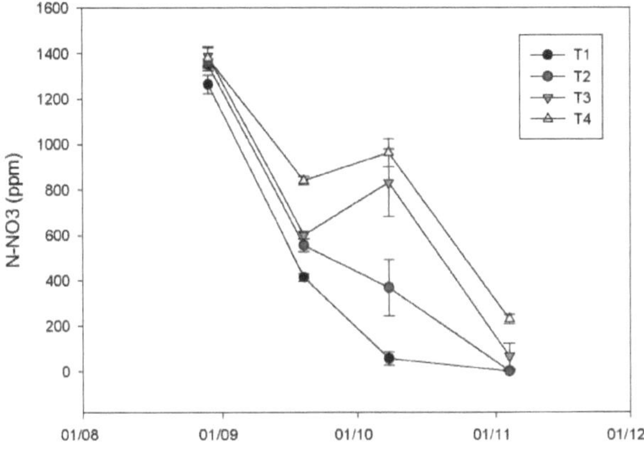

Figure 52. Concentration of N-NO3 (ppm) in sap, in leaves of cauliflower var. Typical in the different treatments. T1, T2, T3 and T4 are the treatments with 84; 130; 190 and 260 kg available N /ha. The vertical bars indicate the standard error.

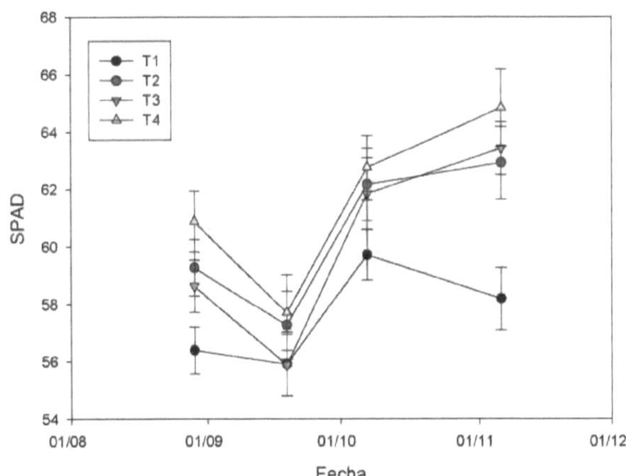

This concentration decreased in the second sampling, after hail, and recovered in the third sampling, after the second mulch fertilization, with the most fertilized treatments, T3 and T4, being significantly higher than the least fertilized, T1 and T2. In the fourth sampling, at the beginning of harvest, the concentration decreased in all treatments below 250 ppm.

SPAD sensor

Figura 53. Chlorophyll content in leaves of cauliflower var. Typical, SPAD units, in the different treatments. T1, T2, T3 and T4 are the treatments, with 84; 130; 190 and 260 kg available N /ha. The vertical bars indicate the standard error.

In the first sampling carried out with the SPAD sensor, significant differences were found before the first mulch fertilization, with the unfertilized treatment T1 being significantly lower. The evolution of its values was similar to that of nitrate in sap. Significant differences were also found in the fourth sampling, between treatment T1 and the rest at the beginning of harvest (Figure 53).

DUALEX sensor

In the measurements made with the DUALEX sensor, significant differences were found for the Chl and NBI indices between treatments on all dates from the first sampling, prior to cover crop fertilization. The evolution of Chl values has been similar in all treatments, except in treatment T3, which on the second sampling date presented lower values than those of treatment T2. The NBI index maintained its values in treatments T3 and T4 and decreased throughout the crop in treatments T1 and T2 (Figures 54 and 55).

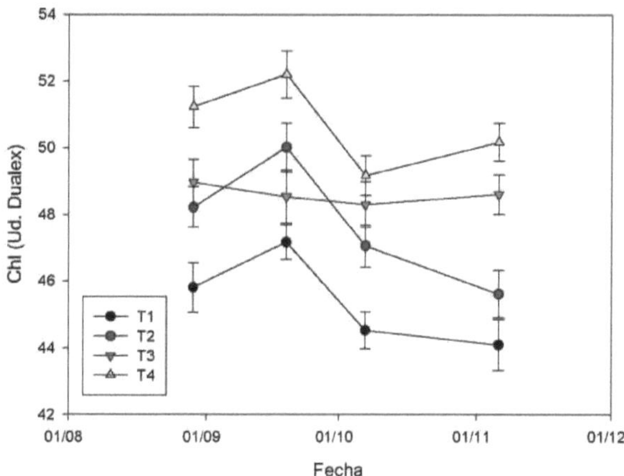

Figura 54. Chl index of Dualex, in cauliflower cv. Typical, in the different treatments of available nitrogen. T1, T2, T3 and T4 are the treatments, with 84; 130; 190 and 260 kg available N /ha. The vertical bars indicate the standard error.

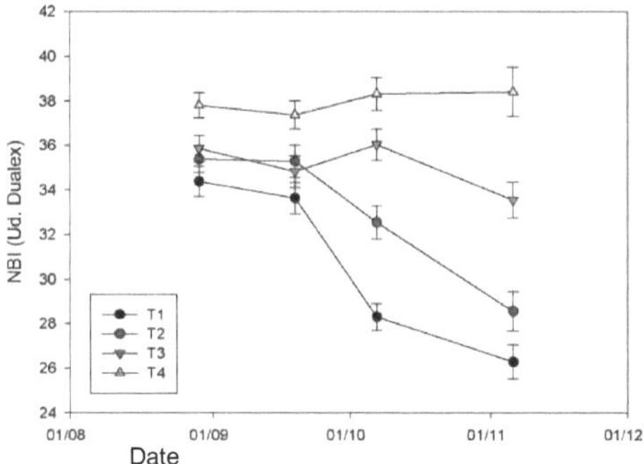

Figura 55. NBI index of Dualex, in cauliflower cv. Typical, in the different treatments of available nitrogen. T1, T2, T3 and T4 are the treatments with 84; 130; 190 and 260 kg available N/ha. The vertical bars indicate the standard error.

MULTIPLEX sensor

The evolution of the SFR and NBI indices of the MULTIPLEX sensor are shown in Figures 56 and 57. Significant differences exist for each date in both

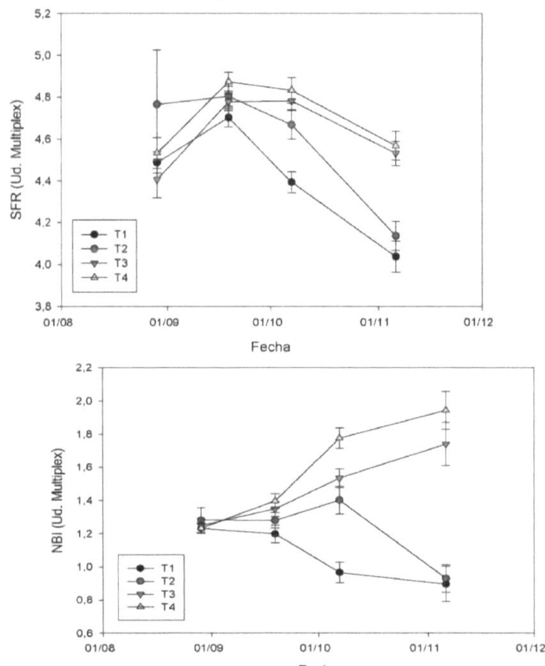

SFR and NBI indices, starting from the first cover fertilization (third sampling). The evolution of the SFR index values was similar in all treatments, with lower values in the least fertilized treatments T1 and T2. Seventeen days after the first mulch fertilization (third sampling), the

values of NBI began to diverge, separating treatments T1 and T2 from treatments T3 and T4.

Figura 56. fodice SFR of Multiplex, in cauliflower cv. Typical, in the different treatments of available nitrogen. T1, T2, T3 and T4 are the treatments, with 84; 130; 190 and 260 kg available N /ha. The vertical bars indicate the standard error.

Figura 57. fodice NBI of Multiplex, in cauliflower cv. Typical, in the different treatments of available nitrogen. T1, T2, T3 and T4 are the treatments, with 84; 130; 190 and 260 kg available N /ha. The vertical bars indicate the standard error.

CROP CIRCLE sensor

In the trajectories performed with the Crop Circle sensor, significant differences were obtained for the NDRE and NDVI indices among the treatments, for each sampling date. In the third sampling, after the first top dressing fertilization, the NDVI and NDRE values of the treatments were ordered according to the available nitrogen (Figures 58 and 59).

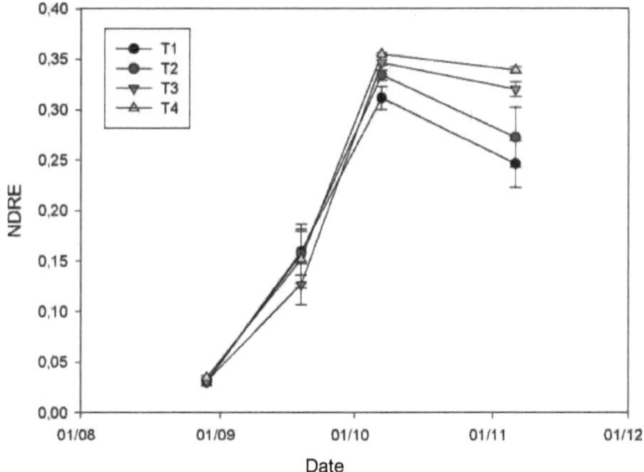

Figura 58. NDRE index of Crop Circle, in cauliflower cv. Typical, in the different treatments of available nitrogen. T1, T2, T3 and T4 are the treatments, with 84; 130; 190 and 260 kg available N /ha. The vertical bars indicate the standard error.

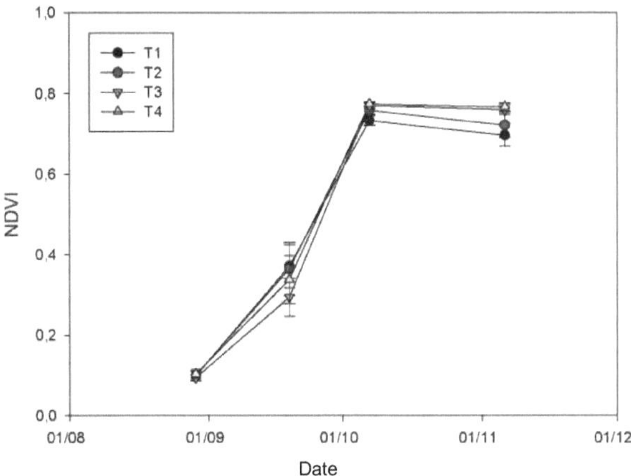

Figura 59. NDVI index of Crop Circle, in cauliflower cv. Typical, in the different treatments of available nitrogen. T1, T2, T3 and T4 are the treatments, with 84; 130; 190 and 260 kg available N /ha. The vertical bars indicate the standard error.

Nitrogen balance

No significant differences were found in the balance and efficiency of nitrogen use (Table 34). High values of nitrogen leached were obtained, as a consequence of the accumulation of rainfall during the trial (219 mm). The negative balance, in all treatments, indicates that the outflow of nitrogen in the system was higher than the inflow. This could be due to an underestimation of the mineralization of organic matter and/or to a lower nitrogen leaching from the soil.

Table 34. Nitrogen balance (kg/ha) up to 0.6 m depth.

	Nmin ini[1]	Nfert[2]	Nminer[3]	Nmin end[4]	Ncos[5]	Nlix[6]	Balance	EUN[7]
				kgN/ha				kg/kgN
T1	84±2a	0	41	7±1	148±22a	36±3	-65±19	81±14
T2	95±3ab	35	41	6±1	195±14ab	76±12	-107±27	83±15
T3	122±6b	68	41	8±1	249±18b	76±12	-102±12	83±9
T4	175±12c	85	41	9±1	301±14b	52±17	-61±20	75±3
	***	ns			***	ns	ns	ns

1) Initial mineral N. 2) N applied as fertilizer. 3) Mean value of mineralized N from 0 to 0.3 m in the 2013 Barcelona variety trial. 4) Mineralized N at the end of the crop. 5) N extracted at harvest. 6) N leached by irrigation or rain. 7) N use efficiency: kg of marketable crop per kg of N available. Significance: *** (p<0.001); ns: not significant. Different letters in the same column indicate significant differences (p<0.05) in a tukey test.

2) Year 2014 var. Typical

The results obtained with this variety are presented up to the time of harvest, when the flooding of the Ebro River prevented its realization.

Coverage, Height and Crop Biomass

Table 35 shows the results and their significant differences, by treatments, for height, cover and biomass of the crop at the beginning of harvest. The height significantly differentiated the treatments according to the amount of nitrogen available. Cover in the unfertilized treatment was significantly lower than in the other treatments. Biomass differentiated the most fertilized treatment from the unfertilized.

Table 35. Crop cover, height and biomass, on 30/12/2014, at the beginning of harvest.

Treatments	Height (m)		Coverage (%)		Biomass (Mg/ha)	
T1	0,49 ± 0,01	a	58 ± 2	a	5,37 ± 0,59	a
T2	0,61 ± 0,01	b	75 ± 3	b	7,96 ± 0,53	ab
T3	0,75 ± 0,01	c	84 ± 1	b	7,43 ± 0,63	ab
T4	0,82 ± 0,01	c	82 ± 3	b	8,84 ± 1,10	b

Different letters differ significantly in a Tukey test (p<0.05). ns: no significant differences.

Nitrogen concentration in leaves

In the first sampling, the nitrogen content in cauliflower leaves, the day before top dressing fertilization, was between 4 and 4.5% in all treatments. From the first sampling, nitrogen concentrations decreased until harvest, being greater in the less fertilized treatments, T1 and T2, with significant differences between treatments. Treatments T3 and T4 reached values of 3.5 % at the end of the crop and treatments T1 and T2 reached values of 2.5 % at that time (Figure 60). According to the Greenwood (1986) model, it was treatments T1 and T2 that had nitrogen concentrations below the values of cntic nitrogen (Figure 61).

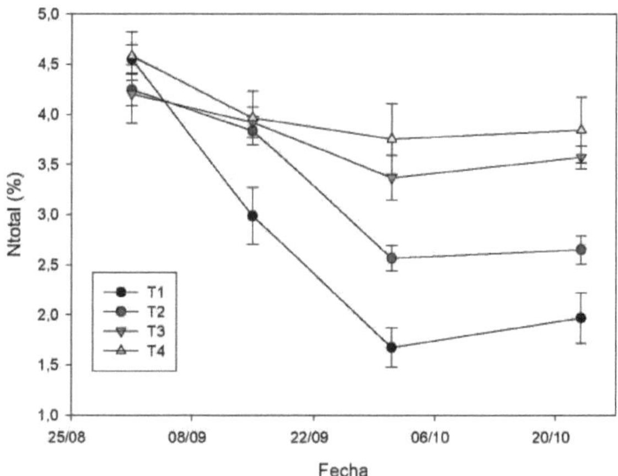

Figura 60. Total nitrogen concentration (%) of cauliflower var. Typical in leaves, throughout the crop in 2014. T1, T2, T3 and T4 are the treatments, with 65; 130; 190 and 260 kg available N /ha.

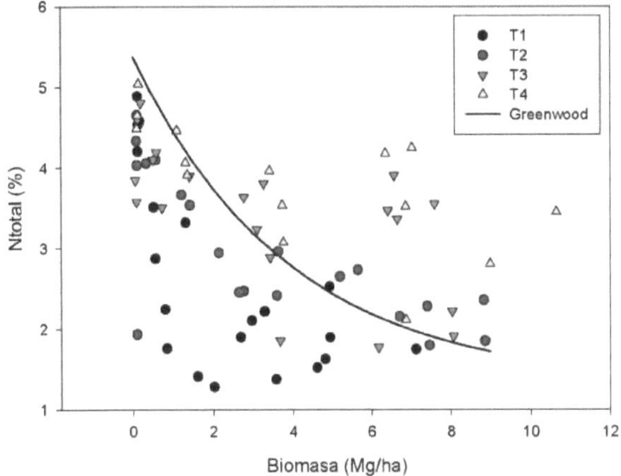

Figura 61. Total nitrogen concentration (%) of cauliflower var. Typical as a function of biomass (Mg/ha) in 2014. T1, T2, T3 and T4 are the treatments, with 65; 130; 190 and 260 kg available N /ha. The analytical curve of the Greenwood 1986 model is presented.

Soil nitrogen content

The initial Nmin content (August 6) in the soil profile up to 0.6 m depth was between 60 and 140 kg/ha (Figure 62). Since the top dressing fertilization, the Nmin content in the soil profile decreased until harvest. At the end of harvest, Nmin decreased in all treatments, below 50 kg/ha. The surface horizon up to 0.15 m appears almost depleted.

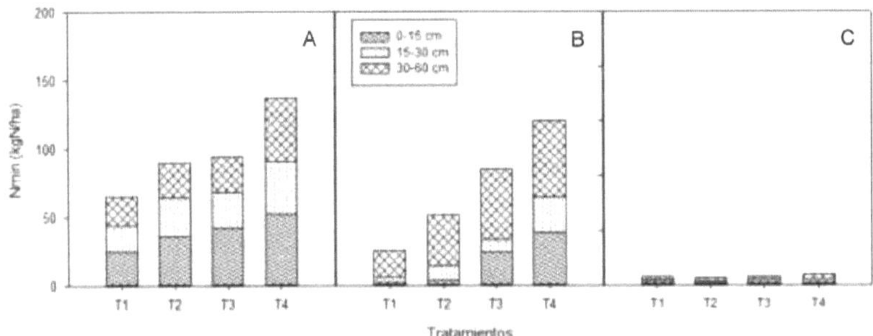

Figure 62. Mineral nitrogen (Nmin) in the soil, from 0 to 60 cm, in the var. Typical, in the anus. 2014, (A) Aug. 6, at the time of transplantation, (B) Sept. 15, after the first
(C) October 28, after the start of pellet formation.

N-Nitrate content in sap

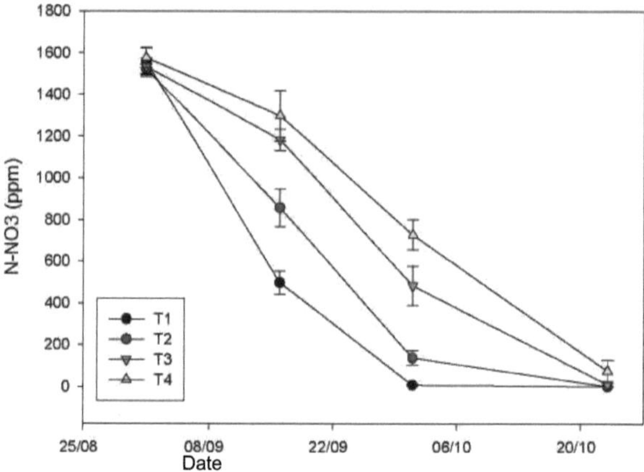

Figure 63. Concentration of N-NO3 (ppm) in sap, in leaves of cauliflower var. Typical in the different treatments. T1, T2, T3 and T4 are the treatments with 65; 130; 190 and 260 kg available N /ha. The vertical bars indicate the standard error.

In the first sampling, one day before top dressing, the concentration of N-NO3⁻ in sap ranged between 1,400 and 1,600 ppm, with no significant differences between treatments. This concentration decreased until the third sampling, at the beginning of pellet formation, values below 800 ppm were found, with significant differences between treatments. In the fourth sampling, nitrate concentration continued to decrease to values below 200 ppm at the beginning of the collection (Figure 63).

SPAD sensor

From the first sampling with the SPAD sensor, after the first mulch fertilization, significant differences were found on all dates. The most fertilized treatments obtained the highest values, and the unfertilized treatments the lowest (Figure 64).

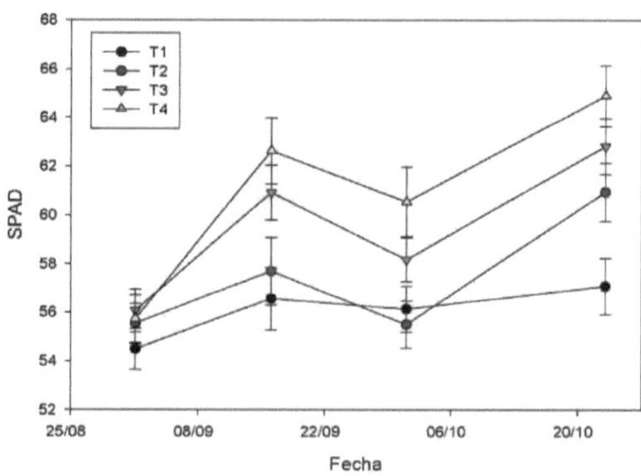

Figura 64. Chlorophyll content in leaves of cauliflower var. Typical, SPAD units, in the different

treatments. T1, T2, T3 and T4 are the treatments, with 65; 130; 190 and 260 kg available N/ha. The vertical bars indicate the standard error.

DUALEX sensor

In the first sampling carried out with the DUALEX sensor, before the first mulch fertilization, and on the rest of the dates, significant differences were found for the Chl and NBI indices. The T4 treatment, the most fertilized, showed the highest values of these indices and the T1, unfertilized, the lowest (Figures 65 and 66).

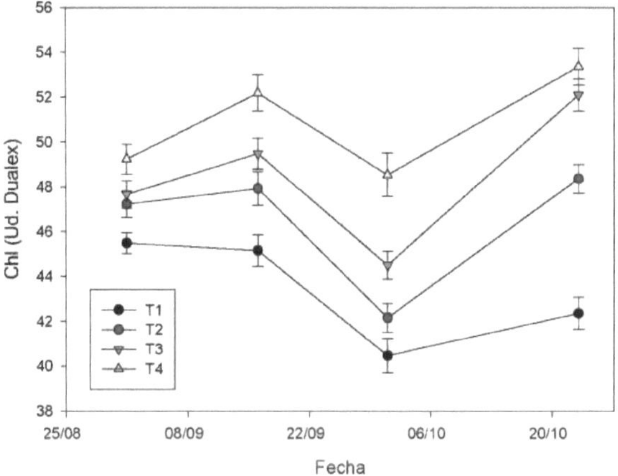

Figura 65. Chl index of Dualex in leaves of cauliflower var. Typical, in the different treatments of available nitrogen. T1, T2, T3 and T4 are the treatments with 65; 130; 190 and 260 kg available N /ha. The vertical bars indicate the standard error.

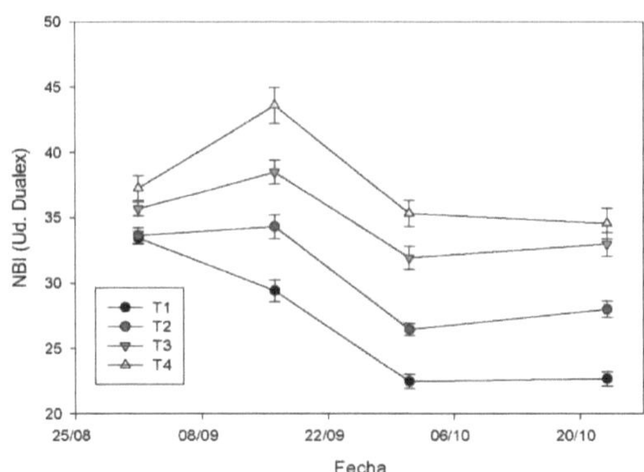

Figura 66. Nitrogen balance index NBI of Dualex in leaves of cauliflower var. Typical, in the different treatments of available nitrogen. T1, T2, T3 and T4 are the treatments, with 65; 130; 190 and 260 kg

available N /ha. The vertical bars indicate the standard error.

MULTIPLEX sensor

The evolution of the SFR and NBI indices of the MULTIPLEX sensor is shown in Figures 67 and 68. From the first sampling, after the first cover crop fertilization, there are significant differences for each date with the NBI index and the SFR index. The evolution of the values of these indexes was similar in all the treatments, with the lowest values for

The evolution of the values of these indexes was similar in all treatments, with the lowest values in the least fertilized treatments T1 and T2.

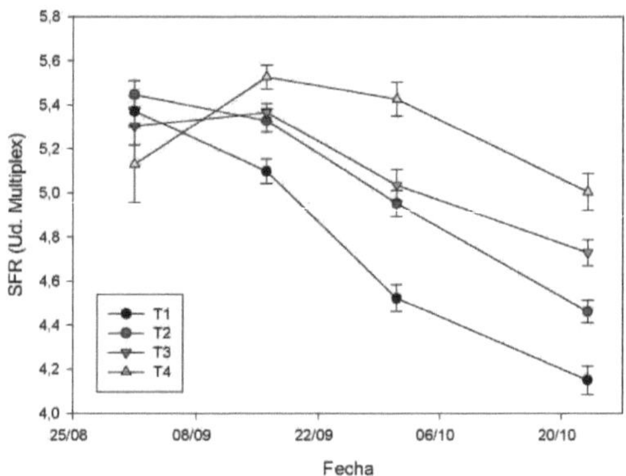

Figura 67.   Multiplex SFR index in leaves of cauliflower var. Typical, in the different treatments of available nitrogen. T1, T2, T3 and T4 are the treatments with 65; 130; 190 and 260 kg available N /ha. The vertical bars indicate the standard error.

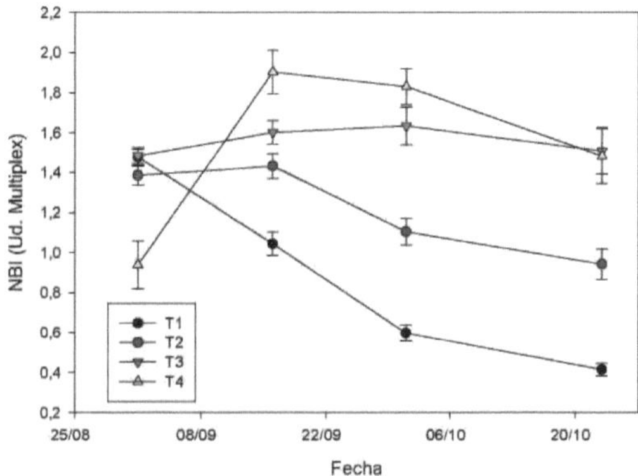

Figura 68.   Multiplex NBI index in leaves of cauliflower var. Typical, in the different treatments of available nitrogen. T1, T2, T3 and T4 are the treatments with 65; 130; 190 and 260 kg available N/ha.

The vertical bars indicate the standard error.

CROP CIRCLE sensor

In the trajectories performed with the Crop Circle sensor, significant differences were obtained for the NDRE and NDVI indices between treatments, for each sampling date The highest values of these indices were obtained for the most fertilized treatments and the lowest for the least fertilized ones (Figures 69 y 70).

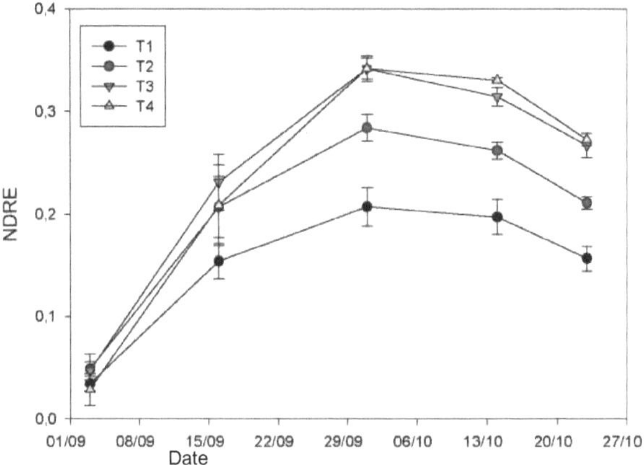

Figura 69. NDRE index of Crop Circle in a cauliflower crop var. Typical, in the different treatments of available nitrogen. T1, T2, T3 and T4 are the treatments, with 65; 130; 190 and 260 kg available N /ha. The vertical bars indicate the standard error.

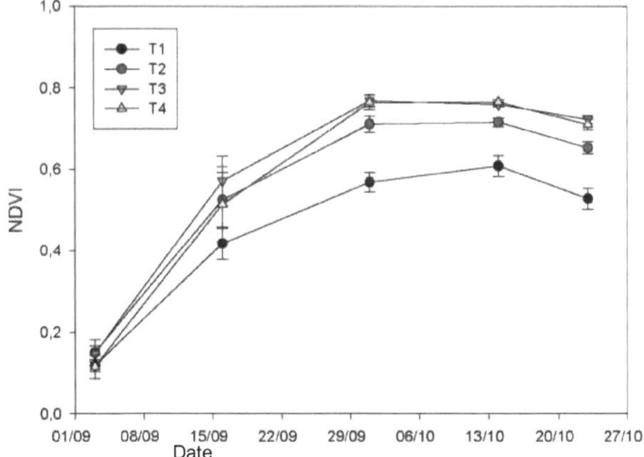

Figura 70. NDVI index of Crop Circle in a cauliflower crop var. Typical, in the different treatments of available nitrogen. T1, T2, T3 and T4 are the treatments, with 65; 130; 190 and 260 kg available N /ha. The vertical bars indicate the standard error.

4.6. Year 2012. Var. Casper

Total production

Possibly due to the high available nitrogen content of the soil, even in the unfertilized treatment (259 kgN/ha), no significant differences were found between treatments in pellet yield (Table 36).

Table 36. Total, leaf and pellet production (kg/ha) of the Casper variety and available nitrogen in the 2012 trial.

Treatments	Navailable	Pellas	Sheets	Total
	kg/ha			
T1	259	34.263	64.497	98.760
T2	359	33.287	61.073	94.360
T3	432	34.949	64.683	99.632
T4	524	33.855	62.165	96.019
		ns	ns	ns

ns: not significant in the analysis of variance.

Nitrogen concentration in leaves

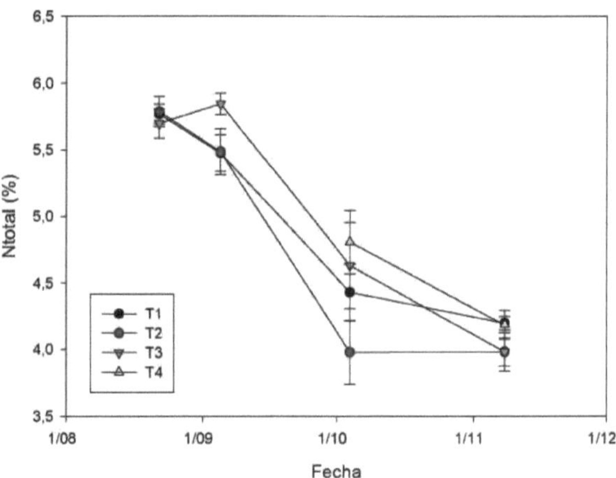

Figure 71. Evolution of nitrogen concentration (%) in cauliflower plant var. Casper as a function of available Nmin in the soil. T1: 259 kgN/ha; T2: 359 kgN/ha; T3: 432 kgN/ha; T4: 524 kgN/ha.

Figure 71 shows the evolution of the nitrogen content (%) in leaves throughout the crop. In the first sampling, twenty days after transplanting, leaf N concentrations are around 6% and decrease as crop biomass increases. No significant differences were found on the different dates, although the treatment with 259 kgN/ha showed the lowest values at the end of the crop. In the fourth sampling, at harvest, very similar values were obtained between treatments, around 4%.

Soil nitrogen content

The Nmin content at the beginning of the crop, up to 0.6 m depth, was between 259 and 424 kg/ha. At the end of harvesting, Nmin decreased in all treatments to values between 14 and 36 kg/ha. The surface horizon up to 0.15 m was almost depleted in all treatments (Figure 72).

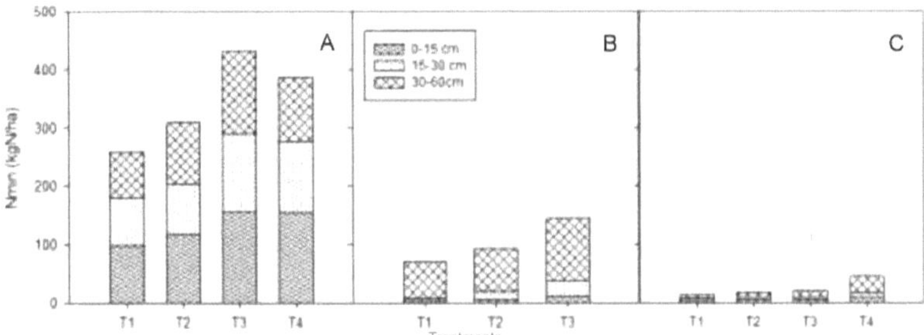

Figure 72. Mineral nitrogen (Nmin) in the soil, from 0 to 60 cm, in a crop of cauliflower var. Casper, in 2012, (A) July 23, at the time of transplantation, (B) October 3, 26 d^as.
after mulching and the beginning of harvesting and (C) November 12, at the end of harvesting.

N-Nitrate content in sap

In the second sampling, fifteen days before top dressing, the concentration of N-NO3⁻ in sap ranged between 5,000 and 6,000 ppm, with no significant differences between treatments. This concentration decreased until harvest, being the concentration in the 259 kgN/ha treatment lower than the rest, but all of them were above the critical levels indicated by Kubota *et al.* (1997) for each date (Figure 73).

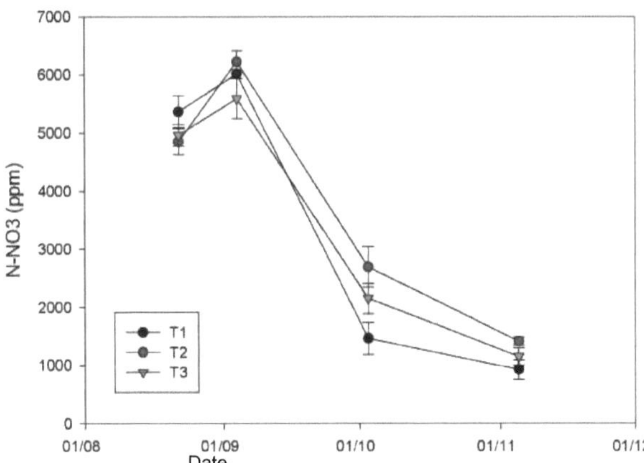

Figure 73. Concentration of N-NO3⁻ (ppm) in sap, in leaves of cauliflower var. Casper in the different treatments as a function of available nitrogen in the soil. T1: 259 kgN/ha; T2: 359 kgN/ha; T3: 432 kgN/ha. The vertical bars indicate the standard error.

Nitrogen balance

The results of the nitrogen balance indicate that treatment T1, with the lowest available Nmin, has been the most efficient with the lowest N losses. In the rest of the treatments, nitrogen losses could have occurred due to nitrate leaching and/or volatilization of the nitrogen applied as fertilizer, which is common in soils with pH greater than seven, where ammonium nitrosulfate is applied (Table 37).

Table 37. Nitrogen balance (kg/ha) up to 0.6 m depth.

	Nmin ini1	Nfert2	Nmin end3 kgN	Nhojas4 /ha	Ninfl5	Balance	EUN6 kg/kgN
T1	259±13a	0	14±1a	200±5		99±4-54±19a	134±11c
T2	309±12b	50	17±2a	194±2		95±153±15b	93±2b
T3	432±13c	0	20±3a	199±9		93±3120±11c	81±3a
T4	424±22c	100	36±7b	201±5		102±6185±23d	65±4a
	***	**		ns		ns***	***

1) Initial mineral N. 2) N applied as fertilizer. 3) Final mineral N. 4) N extracted at harvest in leaves. 5) N extracted at harvest in pellets. 6) N use efficiency: kg of marketable crop per kg of available N (initial Nmin + Nfertilizer). Significance: ** (p<0.01); ***(p<0.001); ns: not significant. Different letters in the same column indicate significant differences (p<0.05) in a Duncan test.

4.7. Year 2014. Var. Casper

Coverage, Height and Crop Biomass

Table 38 shows the results and their significant differences, by treatments, for height, cover and biomass of the crop at the beginning of harvest. Height significantly differentiated the unfertilized treatment from the most fertilized, T4. Cover showed no significant differences between treatments. Biomass differentiated the most fertilized treatment from the unfertilized treatment.

Table 38. Crop cover, height and biomass, on 11/20/2014, at the beginning of harvest.

Treatments	Height (m)		Coverage (%)		Biomass (Mg/ha)	
T1	0,61 ± 0,02	a	81 ± 2	ns	8,75 ± 0,64	a
T2	0,64 ± 0,01	ab	81 ± 4	ns	9,46 ± 0,42	ab
T3	0,67 ± 0,02	bc	86 ± 1	ns	10,52 ± 0,66	ab
T4	0,71 ± 0,01	c	88 ± 1	ns	11,45 ± 0,60	b

Different letters differ significantly in a Tukey test (p<0.05). ns: no significant differences.

Total production

The average total cauliflower production was around 30,000 kg/ha (Table 39). There were significant differences between treatments, depending on the available nitrogen, so that the production in treatments T3 and T4 was significantly higher than in treatments T1 and T2.

Table 39. Total, leaf and pellet production (kg/ha) of the Casper variety and available nitrogen in the 2014 trial.

Treatments	Navailable	Pellas	Sheets	Total
		kg/ha		
T1	104	26.107 a	55.535 a	81.642 a
T2	134	26.780 a	59,660 ab	86,440 ab
T3	190	30.524 b	64,130 ab	94,654 bc
T4	260	32.253 b	66.780 b	99.033 c
		**	***	*

The nonlinear regression analysis of relative total cauliflower production as a function of available nitrogen (Ndisp = Nmin+Nfertilizer) indicates that production stabilizes for Ndisp values of 143 ± 7 kg Ndisp/ha, value of parameter *a* in the regression model [equation 3] (Figure 74). It is understood that the treatments with a nutritional deficit will be those that present a value of available nitrogen lower than the value at which production stabilizes. In this case, treatments T1 and T2 present values of available nitrogen lower than the level at which production stabilizes. Treatments T3 and T4 are above these levels of available nitrogen.

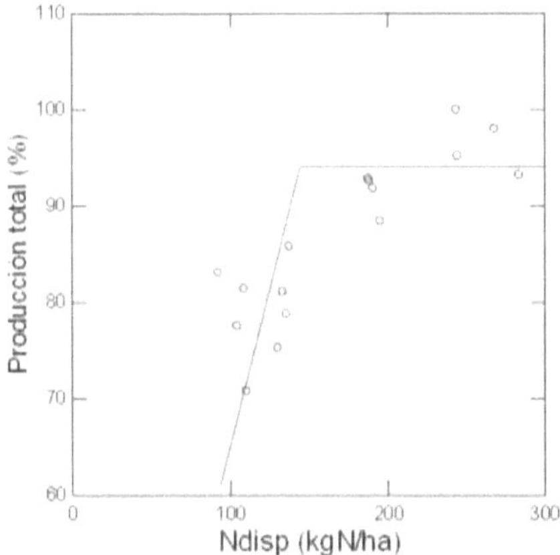

Figura 74. Relative total production of cauliflower pellets var. Casper 2014, as a function of available mineral nitrogen in the soil (Nmin + N fertilizer).

Nitrogen concentration in leaves

In the first sampling, the day before top dressing fertilization, the nitrogen content in cauliflower leaves was between 5 and 5.5% in all treatments. From the first sampling, nitrogen concentrations decreased until harvest, being greater in the less fertilized treatments T1 and T2, and significant differences were observed between treatments. Treatments T3 and T4 reached values of 3 % at the end of the crop and treatments T1 and T2 reached values of 2.5 % at that time (Figure 75). According to the Greenwood (1986) model, it was treatments T1 and T2 that had nitrogen concentrations below the values of cntic nitrogen (Figure 76).

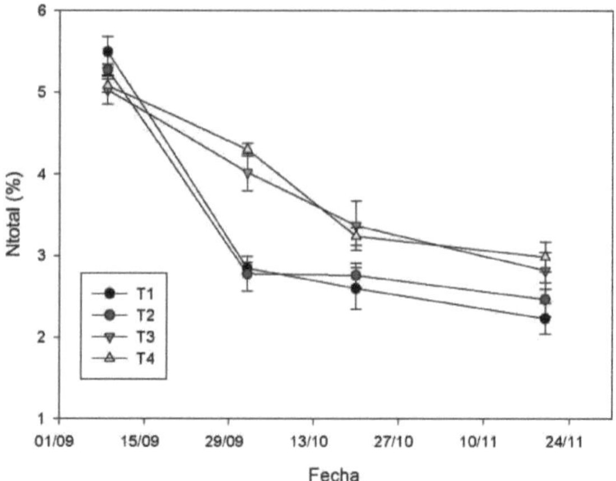

Figura 75. Total nitrogen concentration (%) of cauliflower var. Casper throughout the crop in 2014. T1, T2, T3 and T4 are the treatments, with 104; 134; 190 and 260 kg available N /ha.

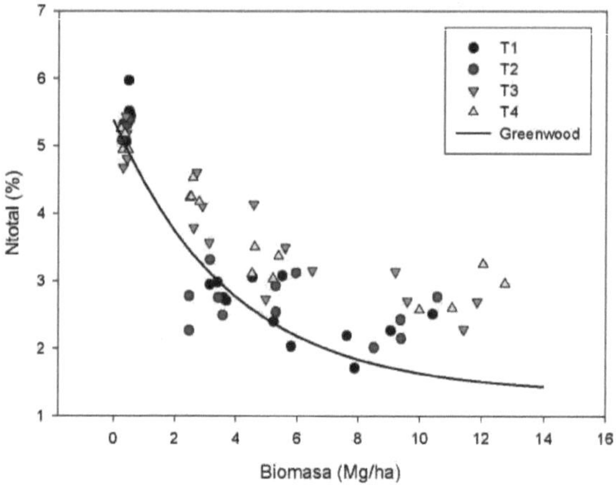

Figura 76. Total nitrogen concentration (%) of cauliflower var. Casper as a function of biomass (Mg/ha) in 2014. T1, T2, T3 and T4 are the treatments, with 104; 134; 190 and 260 kg available N /ha. The analytical curve of the Greenwood 1986 model is presented.

Soil nitrogen content

The initial Nmin content (August 12) in the soil profile up to 0.6 m depth was between 100 and 150 kg/ha (Figure 77). Since the top dressing fertilization, the Nmin content in the soil profile decreased until harvesting. At the end of the harvest, Nmin decreased to below 10 kgN/ha in all treatments. The surface horizon up to 0.15 m appears almost depleted.

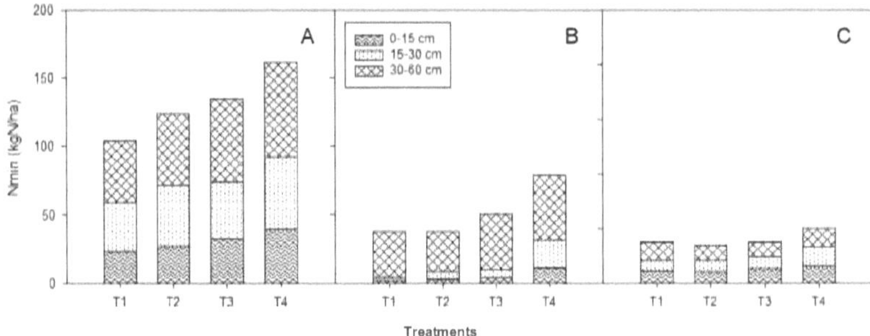

Figura 77. Mineral nitrogen (Nmin) in the soil, from 0 to 60 cm, in the Casper variety, in the year.

2014, (A) August 12, at the time of transplanting, (B) September 29, after fertilization and (C) December 1, at the end of harvest.

N-Nitrate content in sap

In the first sampling, one day before top dressing, the concentration of $N\text{-}NO_3^-$ in sap ranged from 1,000 to 1,200 ppm (Figure 78), with no significant differences between treatments.

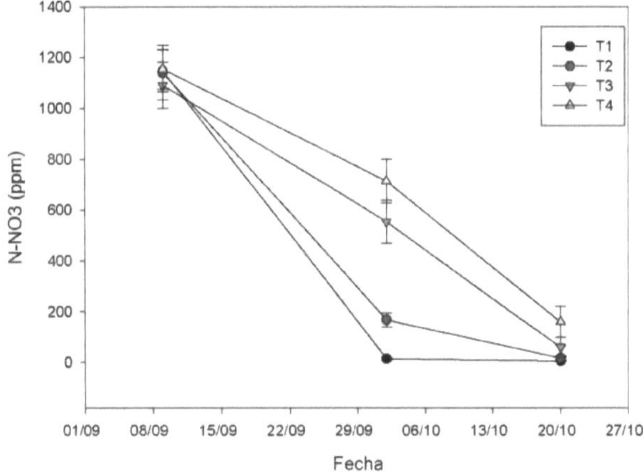

Figura 78. Concentration of $N\text{-}NO_3^-$ (ppm) in sap, in leaves of cauliflower var. Casper in the different treatments. T1, T2, T3 and T4 are the treatments with 104; 134; 190 and 260 kg available N/ha. The vertical bars indicate the standard error.

In the second sampling, eighteen days after the top dressing, this concentration decreased to values below 800 ppm, with significant differences between the most fertilized treatments, T3 and T4, and the least fertilized treatments, T1 and T2. In the third sampling, nitrate concentration continued to decrease to values below 200 ppm at the beginning of the harvest, with significant differences between the most fertilized treatment, T4, and the least fertilized, T1.

SPAD sensor

In the first sampling of the SPAD sensor measurements, before the mulch fertilization, significant differences were found. In the first measurement, the T1 treatment showed the highest SPAD values, and therefore did not indicate lower chlorophyll contents than the other fertilized treatments. In the third sampling, at harvest, the values were not classified according to the nitrogen available in each treatment (Figure 79). Treatment T3 in the third sampling showed values close to those of the unfertilized treatment.

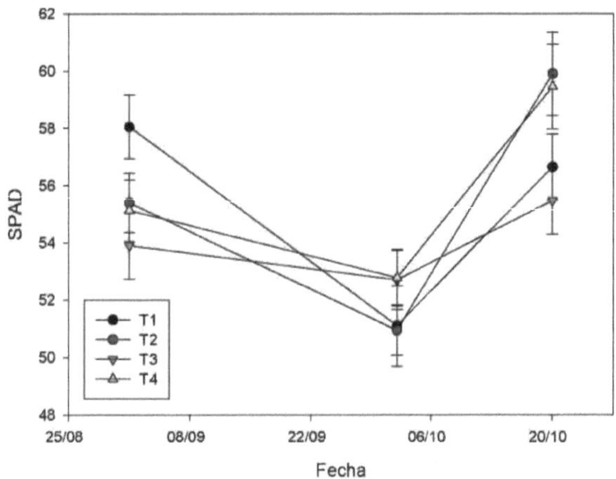

Figura 79. Chlorophyll content in leaves of cauliflower var. Casper, SPAD units, in the different treatments. T1, T2, T3 and T4 are the treatments, with 104; 134; 190 and 260 kg available N /ha. The vertical bars indicate the standard error.

DUALEX sensor

In the first sampling carried out with the DUALEX sensor, before the first mulch fertilization, no significant differences were found for the Chl and NBI indices. In the second sampling, for the Chl index, the T1 treatment was found to have

The T1 and T2 treatments were significantly different from the T3 and T4 in terms of NBI index. In the third sampling, in the days prior to harvest, significant differences were found between treatments, with the most fertilized T4 showing the highest values of these indices and the unfertilized T1 the lowest (Figures 80 and 81).

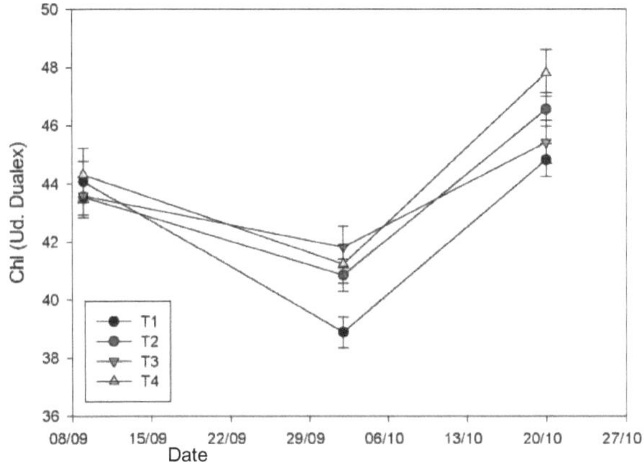

Figura 80. Índice Chl of Dualex in leaves of cauliflower var. Casper, in the different treatments of available nitrogen. T1, T2, T3 and T4 are the treatments, with 104; 134; 190 and 260 kg available N /ha. The vertical bars indicate the standard error.

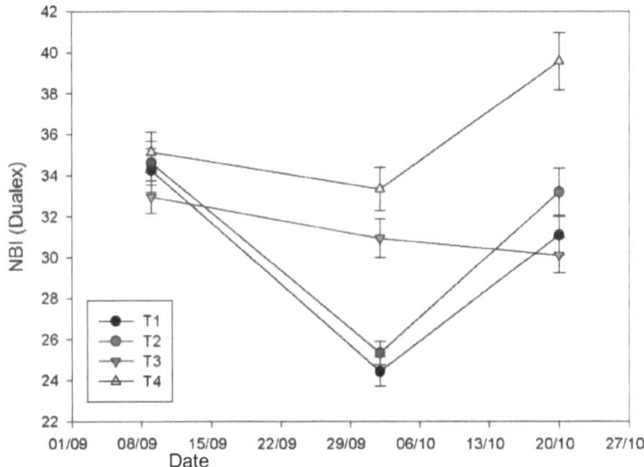

Figura 81. NBI index of Dualex in leaves of cauliflower var. Casper, in the different treatments of available nitrogen. T1, T2, T3 and T4 are the treatments, with 104; 134; 190 and 260 kg available N /ha. The vertical bars indicate the standard error.

Figura 82.

MULTIPLEX sensor

The evolution of the SFR and NBI indices of the MULTIPLEX sensor is shown in Figures 82 and 83. From the first sampling, after cover crop fertilization, there are significant differences for each date with the NBI index and the SFR index.

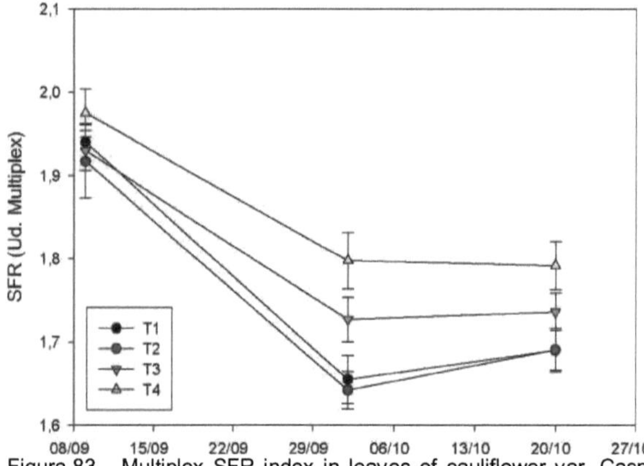

Figura 83. Multiplex SFR index in leaves of cauliflower var. Casper, in the different treatments of available nitrogen. T1, T2, T3 and T4 are the treatments with 104; 134; 190 and 260 kg available N /ha. Vertical bars indicate the standard error.

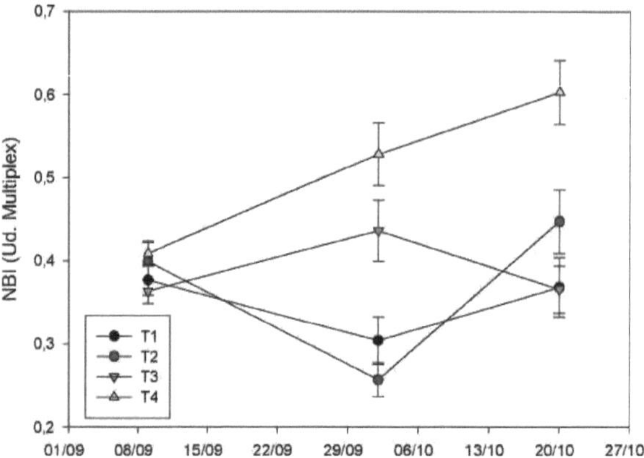

Multiplex IBN in leaves of cauliflower var. Casper in the different treatments of available nitrogen. T1, T2, T3 and T4 are the treatments, with 104; 134; 190 and 260 kg available N /ha. The vertical bars indicate the standard error.

The evolution of the values of these indices was similar in all treatments, with the lowest values being obtained in the least fertilized treatments T1 and T2. Treatment T3 in the third sample, for the NBI index, showed a downward trend that differs from the rest of the treatments, in agreement with what was observed with the Dualex sensor and SPAD.

CROP CIRCLE sensor

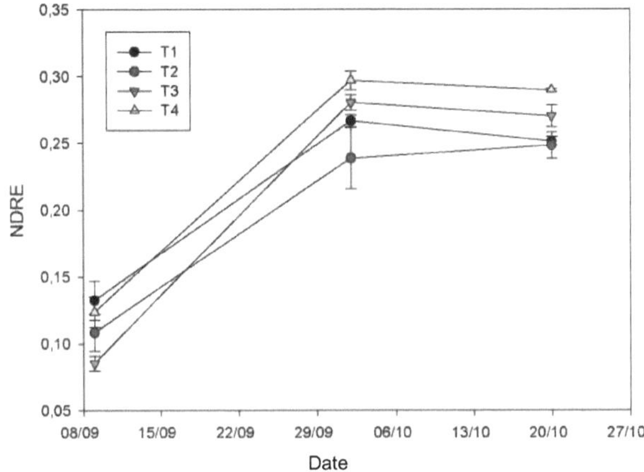

Figura 84.   Crop Circle NDRE index in leaves of cauliflower var. Casper, in the different treatments of available nitrogen. T1, T2, T3 and T4 are the treatments, with 104; 134; 190 and 260 kg available N /ha. The vertical bars indicate the standard error.

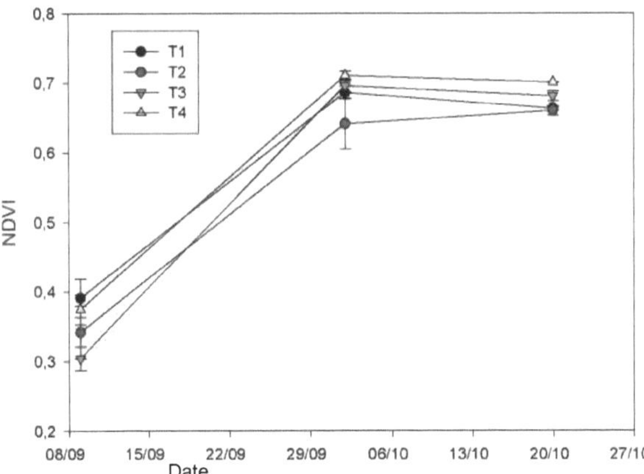

Figura 85.   Crop Circle NDVI index in leaves of cauliflower var. Casper, in the different treatments of available nitrogen. T1, T2, T3 and T4 are the treatments, with 104; 134; 190 and 260 kg available N /ha. The vertical bars indicate the standard error.

In the trajectories performed with the Crop Circle sensor, significant differences were obtained for the NDRE and NDVI indices between treatments, for each sampling date. The highest values of these indices were obtained for the most fertilized treatments and the lowest for the least fertilized treatments (Figures 84 and 85).

Nitrogen balance

The results of the nitrogen balance (Table 40) indicate a negative balance in all treatments, with higher nitrogen outputs from the system than inputs, which could be due to a nitrogen supply through the mineralization of organic matter, not measured in this trial.

Extractions were significantly higher in the most fertilized treatment, T4. In this trial we did not obtain sufficient data to estimate the nitrogen leachate to be able to account for it in the balance sheet.

Table 40. Nitrogen balance (kg/ha) up to 0.6 m depth.

	Nmin ini[1]	Nfert[2]	Nmin end[3]	Ncos[4]	Balance	EUN[5]
			kgN/ha			kg/kgN
T1	104±4a	0	8±1	202±27a	-106 ± 27	253±18a
T2	124±10b	10	7±0	208±20a	-81 ± 19	200±4b
T3	135±55b	55	8±1	261±22ab	-79 ± 23	161±3c
T4	162±98c	98	10±1	286±17b	-36 ± 13	125±6d
	***		ns	*	ns	***

1) Initial mineral N. 2) N applied as fertilizer. 3) Mineral N at the end of the crop. 4) N extracted at harvest. 5) N use efficiency: kg of commercial crop per kg of N available. Significance: *** (p<0.001); * (p<0.05) ns: not significant. Different letters in the same column indicate significant differences (p<0.05) in a tukey test.

5. DISCUSSION

5.1. Harvest and fertilization threshold

In the Barcelona variety trial, in 2012, 192 kg/ha of available nitrogen were necessary to reach the maximum potential pellet yield. The T2 treatment showed similar production levels to the most fertilized treatments (T3 and T4).

In 2013, in the Barcelona variety, 178 kg/ha of available nitrogen were required to achieve a total pellet yield close to 20 t/ha. In 2014,   179 kg/ha of available nitrogen were needed to achieve a total pellet yield of about 20 t/ha.
total harvest over 16 t/ha.

In the Barcelona variety, in the three trials, the mean value of available N above which no yield response was found was 184 ± 20 kgN/ha. In the Typical variety in 2013, this amount was 189 ± 45 kgN/ha. In Casper in 2012, due to the high values of mineral nitrogen available in the soil, no differences in production were obtained and in 2014, for this same variety, the value found was 143 ± 7 kgN/ha, these values were significantly lower than in the rest of the trials. Possibly, the higher harvest values in this variety have been affected by the methodology used in harvesting, in which the destination was not adequately considered and because it is a cauliflower intended for industry and whose average skin weight is higher than that of the varieties intended for fresh consumption, such as Barcelona and Typical.

The values found, with their confidence interval, are close to the optimum nitrogen fertilization for this crop, which according to Everaarts *et al*. (1996), is around 224 kg N/ha. Csizinszky (1996) obtained the maximum yield in green cauliflower with 294 kg N/ha of fertilizer. Rahn *et al*. (1998) determined the production ceiling for values between 240 and 300 kg N/ha of fertilizer. Rather *et al*. (2000) determined the optimum dose at 250 kg N/ha, as the sum of mineral N in the soil (Nmin) at transplanting and the N applied as fertilizer; these authors concluded that when Nmin was higher than 210 kg N/ha in the soil horizon between 0 and 30 cm or higher than 270 kg N/ha in the soil profile from 0 to 90 cm, no response to nitrogen fertilization was found.

5.2. Nitrogen in plant

In the Barcelona variety, in 2012, the nitrogen concentration in cauliflower leaves in the first sampling was higher than 5% in all treatments.

It decreased in all treatments as crop biomass increased, until harvest, when values of 2.5% were obtained for T1, 3.5% for T2 and values higher than 4% for treatments T3 and T4. For this trial, according to Greenwood's model (1986), only treatment T1 had nitrogen concentrations below the cntic nitrogen values, justifying the lower production obtained for this treatment.

In 2013, for the Barcelona variety, the nitrogen content in cauliflower leaves was around 4.5% in the first sampling in all treatments. Subsequently, it maintained values similar to the initial ones, possibly due to the decrease in biomass that occurred after a hail event. At the end of the crop, treatment T1 obtained a value close to 2.5%, T2 3.5% and treatments T3 and T4 values higher than 4.5%. According to Greenwood's model (1986), T1 had nitrogen concentrations below the cntic nitrogen values, thus justifying the lower production values obtained. Treatment T2, which also had a lower production, had total nitrogen values above the cntic curve, but close to the deficit zone differentiated by the model.

In 2014, in the Barcelona variety trial, the nitrogen content in cauliflower leaves in the first sampling, the day before top dressing fertilization, was around 5% in all treatments. Treatments T3 and T4 reached values of 3.5% at the end of the crop and treatments T1 and T2 reached values of 2.5% at that time. According to the Greenwood (1986) model, it was treatments T1 and T2 that had nitrogen concentrations below the cntic nitrogen values.

For the Typical variety, in 2013, the nitrogen content in cauliflower leaves remained around 4.5% in treatments T3 and T4. Subsequently, nitrogen concentration gradually decreased to 3% in treatments T3 and T4 and 2% in treatments T1 and T2, at the beginning of harvesting. According to the Greenwood (1986) model, treatments T1 and T2 have had nitrogen concentrations below the cntic nitrogen values.

In 2014, in the Typical variety, in the first sampling, the nitrogen content in cauliflower leaves was between 4 and 4.5 % in all treatments. Treatments T3 and T4 reached values of 3.5 % at the end of the crop and treatments T1 and T2 reached values of 2.5 % at that time. According to Greenwood's model (1986), it was treatments T1 and T2 that had nitrogen concentrations below the values of cntic nitrogen.

In the Casper variety, in 2012, in the first sampling, leaf N concentrations were around 6% and decreased as crop biomass increased. In the fourth sampling, at harvest, very similar values were obtained between treatments, around 4%. In this trial, no biomass data were taken and therefore the nitrogen curve was not plotted.

In 2014, for the Casper variety, in the first sampling, the nitrogen content in cauliflower leaves was between 5 and 5.5 % in all treatments. Treatments T3 and T4 reached values of 3 % at the end of the crop and treatments T1 and T2 reached values of 2.5 % at that time. According to Greenwood's model (1986), it was treatments T1 and T2 that had nitrogen concentrations below the values of cntic nitrogen.

The nitrogen content in cauliflower was studied by Rincon *et al.* (2001) in Spain, applying a nitrogen fertilization of 325 kg N/ha, and obtaining an average nitrogen percentage 20 days after transplanting close to 6%, reaching values around 5% at the end of the crop. In the experiment on cauliflower by Rincon *et al.* (2001), the concentration of total nitrogen decreased until the end of the crop. As in other trials conducted on cauliflower by Kage *et al.* (2002) and on broccoli in the studies of Magnifico *et al.* (1979) and Rincon *et al.* (1999), where the trend is repeated.

The concentration of total nitrogen in cauliflower leaves has shown a similar behavior in all trials. In the first samples, the total nitrogen concentration was above 4%, with higher initial values in the Casper variety, and later, until harvest, this value decreased in all trials due to the dilution effect in leaves produced by the increase in crop biomass. At the end of the crop, the treatments with less available nitrogen presented lower values of total nitrogen concentration in leaves. In spite of being a destructive and laborious method, it has been able to discriminate the deficit treatments throughout the crop cycle in the different trials.

The relationship between total nitrogen concentration and crop biomass has been studied by several authors such as Greenwood (1986 and 1996) or Rahn *et al.* (2010a and 2010b) and presented in the introduction and in the material and methods section of this work. This relationship has been able to discriminate through the nitrogen curve the treatments considered as deficient, which were found with a level of available nitrogen below the value for which no response in crop production was obtained.

This discrimination was more sensitive from biomasses higher than 1 Mg/ha, below which the concentration of cntic nitrogen is independent of aerial biomass according to studies cited by Justes *et al.* (1994).

5.3. Nitrogen in the soil

The analysis of the mineral nitrogen present in the first 60 cm of the soil showed in all the trials the surface horizon exhausted at the end of the crop. In spite of not finding significant differences due to the variability present in the soil itself, it has grouped and identified the treatments according to the mineral nitrogen available in the soil throughout the crop.

Although it is a laborious method of analysis, the analysis of mineral nitrogen in the soil has made it possible to know the amount of mineral, nitric and ammoniacal nitrogen available to the crop at the times when the samples were taken. This soil analysis can be carried out at different times during cultivation: 1) at the beginning of cultivation at a certain soil depth, for example, in the case of the Nmin system (Feller and Fink, 2002), and 2) to determine the top dressing fertilizer based on the soil analysis of the first 30 cm and, depending on the values obtained, to diagnose whether or not the fertilizer can be expected to increase production (Krusekopf *et al.*, 2002). These systems are very useful in the case of horticultural crops because it is very common that the residual Nmin values in the soil from the previous crop are very high (Ramos *et al.*, 2002; Vazquez *et al.*, 2006) and, therefore, fertilization can be greatly reduced or even eliminated.

Delaying the soil analysis until moments before the top dressing fertilization, allows us to adjust the nitrogen fertilization more efficiently, since we take into account the possible mineralization of the soil organic matter from the beginning of the crop, as well as the possible losses by leaching. This allows us to adjust the dosage more precisely.

The main drawback of these systems is the cost of soil sampling and analysis and the high spatial variability of nitrate in soil (Lopez-Granados *et al.*, 2002; Giebel *et al.*, 2006). Simple and inexpensive methods have been developed to reduce the cost of analysis (Hartz, 1994; Sepulveda *et al.*, 2003; Thompson *et al.*, 2009).

5.4. Mineralization

In the Barcelona variety trial, in the years 2012 and 2013, soil organic matter mineralization measured in the field reached an average value of 41 kg N/ha for the topsoil crop, extrapolated to 0.3 m depth.

In the 2014 trial, for the same variety, the mineralization reached an average value of 57 kg N/ha for the crop canopy, extrapolated to the topsoil layer up to 0.3 m.

In systems based on soil measurements, simulation models or N balance, the N input from mineralization of soil organic matter can be an important part of the N available to crops and, therefore, its determination is of interest. Fink and Scharpf (2000) and Tremblay et *al.* (2001) in work using horticultural plants estimated a mineralization rate of 5 kg N/ha and week, which implies that in a 90-day cauliflower crop, an N supply of 60 kg N/ha can be achieved, which are close to the values obtained in our trials.

Nitrogen extractions for this crop, according to several studies, can vary between 150 and 300 kg N/ha (Everaarts *et al.*, 1996), 170 and 250 kg N/ha (Everaarts, 2000) and 250 and 498 kg N/ha (Vazquez *et al.*, 2010). In our trials, the average extractions (leaves and pellets) of cauliflower, in the treatments considered non-deficit, were 246 kg N/ha. Taking into account the average mineralization obtained, 46 kg N/ha, the nitrogen supply through the mineralization of soil organic matter could account for about 20% of the nitrogen extractions of the plant in our trials.

5.5. Nitrogen balances

In the deficit treatments, the available nitrogen was not sufficient to cover its needs, presenting a lower nitrogen extraction by the crop.

Mineralization was a relatively important source of nitrogen, depending on soil fertility. The average mineralization rate in the Valdegon trials reached 41 kg N/ha in the years 2012 and 2013, and 57 kg N/ha in the year 2014, for the growing season in the top 0.3 m layer.

Significant losses, which have not been evaluated, may occur due to volatilization of nitrogen applied as fertilizer (parameter not measured), which is common in soils with low pH where ammonium nitrosulfate is applied as fertilizer (Meisinger and Randall, 1991).

The estimated leaching in almost all the trials has not been very important, perhaps due to good irrigation scheduling. However, in the trial of the Typical variety, in 2013, leaching values of 76 kg N/ha were reached in some treatments, which may be due to the accumulation of rainfall in several periods (219 mm).

The average extractions (leaves and pellets) of cauliflower in the trials carried out, in the treatments considered as non-deficit (those with available nitrogen values in the soil above the value for which no production response was obtained), were 246 kg/ha. The average available nitrogen (Nmin + Nfertilizer) was 190 kg/ha, and the average mineralization was 46 kg N/ha, very close to the values found by Tremblay *et al.* (2001) in a 90-day cauliflower crop.

To cover the needs of 246 kg/ha and without considering fertilizer efficiency and nitrogen losses, 270 kg N/ha were applied (available N + mineralization). Rather *et al.* (2000), determined the optimum dose to be 250 kg N/ha, as the sum of mineral N in the soil (Nmin) at transplanting and the N applied as fertilizer.

The average Navailable N threshold seen above for Barcelona and Typical is around 186 kg N/ha, if we add the average mineralization measured, 46 kg N/ha, it makes a total of 232 kg N/ha, a value very close to the average needs of 246 kg/ha.

It can be concluded in this section that the results of the analysis of the balance in the different trials confirm the usefulness of the Nmin method for the design of nitrogen fertilization of cauliflower, as well as the importance that mineralized nitrogen can acquire in the balance and the need to reduce leaching losses through correct irrigation scheduling. The average extractions of the cauliflower varieties studied, in the treatments considered as non-deficit, were 246 kg of nitrogen per hectare.

5.6. SPAD sensor

The results obtained with the SPAD sensor were highly variable and of low repeatability. In general, no differences between treatments were detected before top dressing fertilization, i.e. up to about thirty days after transplanting. Subsequently, only on one occasion was SPAD able to detect differences within twenty days of fertilizer application. SPAD was able to detect differences in the later determinations, close to pellet formation.

SPAD sampling is fast, although being a method of measurement by transmittance (leaf pinching) it is affected by several factors, e.g. time of day, leaf surface moisture, leaf thickness or position, etc., which require a strict protocol for its determination (Hoel and Solhaug, 1998; Martinez and Guiamet, 2004). Some works show that the chlorophyll meter is only capable of detecting severe nitrogen deficiencies (Villeneuve *et al.*, 2002), or requires a well fertilized plot as a reference (Westcott and Wraith, 1995). Other authors question its use to reliably assess plant N concentration (Himelrick *et al.*, 1993). (1993) on bell pepper, Goffart *et al.*, (2006) on endive and Gianquinto *et al.*, (2003) on potato recommend its use, while Villeneuve *et al.*, (2002) on broccoli or Tremblay *et al.*, (2002) on green bean recommend its use on cabbage.

are not recommended for subscriber recommendation.

5.7. DUALEX sensor

The initial Chl values of Dualex in the Barcelona variety were around 44 units (Table 41). In the final determination, the values in the most fertilized treatments (T3 and T4) maintained the same value or increased up to 9%, while they decreased by 2.3% in the least fertilized treatments (T1 and T2). The NBI index started from initial values close to 4 units, which increased slightly (2%) at the end of the measurements in the most fertilized treatments and decreased radically in the least fertilized treatments (20.7% on average). It seems therefore that, throughout the crop, the Chl and NBI indices maintained or slightly increased their values in the fertilized treatments while they decreased in the unfertilized ones, with a gradation between T1 and T4. This decrease has been very evident with the NBI index.

Table 41. Mean initial and final values of Dualex, of fertilized (F) and non-fertilized (NF) treatments, in the Barcelona variety trials in the years 2012, 2013 and 2014, with the percentage of variation (±A).

	Initial Chl	Final Chl	±Д	Initial NBI	Final NBI	±Д
Dualex F	46,2	50,3	8,9%	35,0	35,7	2,0%
Dualex NF	43,5	42,5	-2,3%	33,0	26,2	-20,7%

In the Typical variety (Table 42), the initial Chl values of Dualex were close to 48 units and increased slightly (3.5%) in the fertilized treatments while they decreased (-3.4%) in the unfertilized treatments. From an initial value of 35 units, the BIL decreased in both fertilized and unfertilized treatments, but in the latter it did so to a greater extent (-22.9%).

Table 42. Mean initial and final values of Dualex, of fertilized (F) and non-fertilized (NF) treatments, in the Typical variety trials in the years 2013 and 2014, with the percentage of variation (±Д).

	Initial Chl	Final Chl	±Д	Initial NBI	Final NBI	±Д
Dualex F	49,3	51,1	3,5%	36,4	34,9	-4,3%
Dualex NF	46,7	45,1	-3,4%	34,2	26,4	-22,9%

In both varieties, chlorophyll content is maintained or slightly increased in the fertilized treatments and decreased in the unfertilized treatments. The NBI index is shown to be more discriminating due to the introduction of the stress factor provided by the flavonoid index used in the calculation. This trend of the NBI index coincides with that observed in the trials of Padilla *et al.* (2014) on melon and Cerovic *et al.* (2015) on grapevine.

The results obtained with the DUALEX sensor indicate that, in general, it has been more repeatable and sensitive than SPAD in detecting differences between treatments before top dressing fertilization and within twenty days of fertilizer application, which could be useful for correcting possible nitrogen fertilization deficiencies depending on the length of the crop cycle. The DUALEX sensor provides two measures related to chlorophyll content Chl and nitrogen balance NBI. Although it appears that the NBI index is more discriminating than the Chl index, the use of both allows a better diagnosis of the plant status in relation to nitrogen fertilization. Sampling with DUALEX is fast, and also, being a transmittance measurement method (leaf pinch) it is affected by the same factors as SPAD.

5.7.1. Dualex Critical NBI Curve

After analyzing the results obtained in the trials with the Dualex sensor, in which it showed sensitivity to significantly differentiate the deficit treatments, it was observed that there was a significant linear correlation between the values of the Dualex NBI index and the concentration of total nitrogen in the leaf during the development of the crop. Padilla *et al.* (2014) in melon and Cerovic *et al.* (2015) in grapevine found high correlations between the NBI index and nitrogen concentration in the plant leaf.

Figure 86 shows the values obtained for the Dualex NBI index and total leaf nitrogen concentration for the Barcelona variety in the 2012, 2013 and 2014 trials.

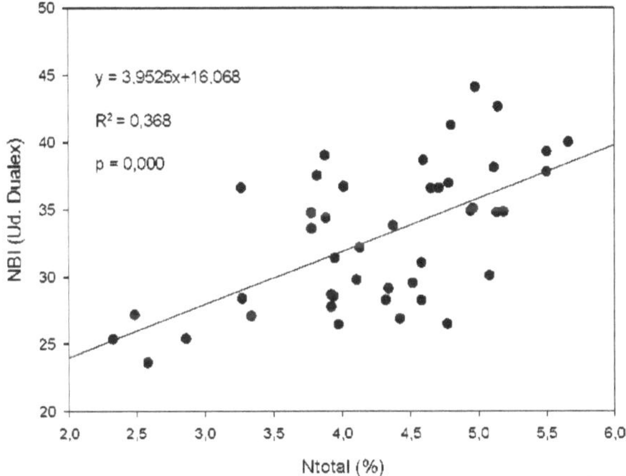

Figure 86. Relationship between Dualex NBI index and total nitrogen concentration in cauliflower leaf of the Barcelona variety during the years 2012, 2013 and 2014. The p-value indicates the degree of statistical significance.

Figure 87 shows the values for the Typical variety for the years 2013 and 2014.

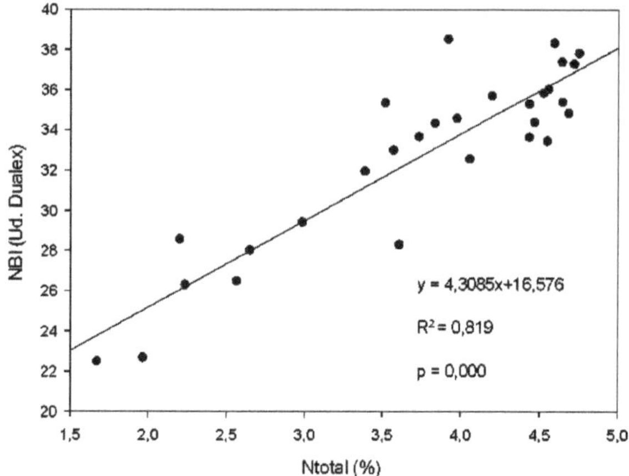

Figura 87. Relationship between Dualex NBI index and total nitrogen concentration in cauliflower leaves of the Typical variety during 2013 and 2014. The p-value indicates the degree of statistical significance.

A comparison of the regression parameters for both varieties was made and it was found that there were no significant differences between them. Therefore, it was possible to establish a common regression for both varieties, the result of which is shown in Figure 88.

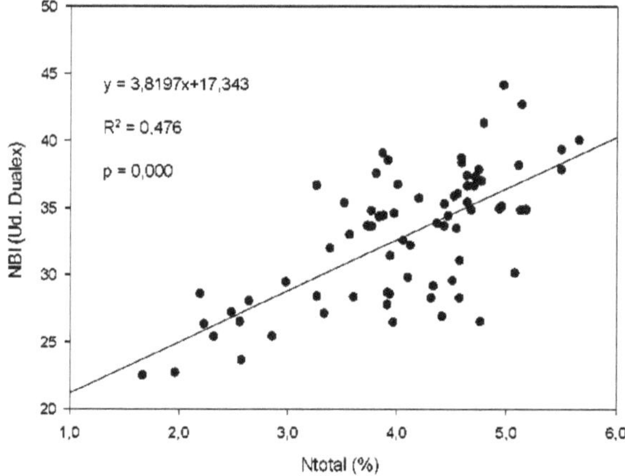

Figura 88. Relationship between Dualex NBI index and total nitrogen concentration in cauliflower leaves of Barcelona and Typical varieties. The p-value indicates the degree of statistical significance.

Once the existence of a highly significant linear relationship between the two parameters was verified, and given that total nitrogen concentration is related to biomass by means of the cntic nitrogen functions, a function of BNB and crop biomass was fitted. This function has a similar interpretation to that of the cntic nitrogen model and is intended to delimit a cntic BN region.

For the construction of the cntic curve of the Dualex NBI index, the results of biomass and the Dualex NBI index were used. The methodology indicated by Olasolo (2013) was followed to fit the analytical curve of nitrogen in jud^a, and for this purpose, the treatments with the lowest values were selected from among the significantly more productive treatments. With these data, a non-linear model was fitted, derived from Greenwood (1986), and equation:

$$y = a * (1 + e *)^{-\hat{a}x} \qquad [8]$$

where "y" is the value of the NBI index (ud. Dualex) and "x" is the crop biomass (Mg/ha). A model was fitted for the shortest cycle variety, Barcelona, and another model for the medium cycle Typical variety. Both models had highly significant parameters and regressions. It was found that there were no significant differences between the parameters of the two models, and a common model was fitted for the two varieties. The fitted function was as follows:

$$y = 19.815 \, (1 + e^{-0.23x} \qquad)[9]$$

regression and parameters being highly significant (Table 43).

Sum of squares (SC), degrees of freedom (gdl), mean squares (MS) and value of the F statistic. Value of parameters "a" and "b" of the function [8], standard error (SE) and value of the t statistic (tobserved). Significance: *** = p<0.001.

Source	SC	gdl	CM	F	Parameter		SE	tobservada
Regression	19.036,60	2	9.518,30	2060,8***				
Error	73,89	16	4,61871		a =	19,815	0,598	33,11***
Total	19.110,50	18			b =	0,23	0,041	5,56***
R² corrected	0,8008							

Figure 89 shows the common model obtained for the two varieties, as well as the upper and lower limits of the confidence interval (95%) of the model. For biomasses greater than 1 Mg/ha, the curve correctly identifies 92% of the most productive, non-nitrogen-deficient treatments and 79% of the nitrogen-deficient treatments. Therefore, this curve determines the threshold values by

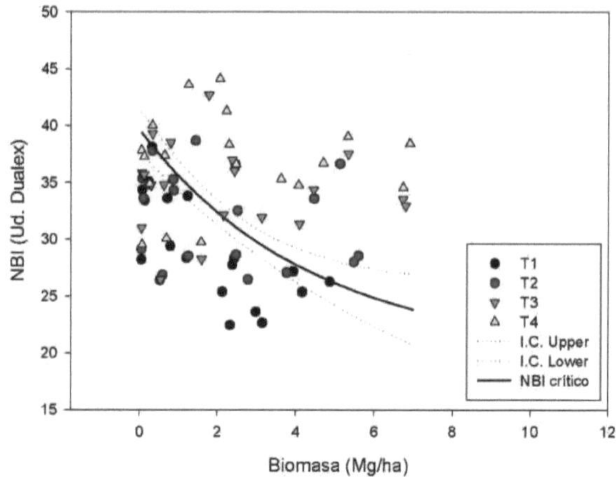

below which the cauliflower crop is in a state of nitrogen nutritional deficit.

Figura 89.   Dualex NBI index and biomass (Mg/ha) of BarcelonandTypical var   .

2012, 2013 and 2014 trials. Adjusted analytical curve of NBI as a function of biomass of the

(NBI cntico) and 95% confidence interval (I.C. Upper and I.C. Lower).

Taking into account the obtained values of the curve, the   interval   of

model confidence and crop biomass, we can propose some momos sufficiency values (Table

44):

Table 44. Minimum sufficiency values for the NBI index of Dualex as a function of biomass, in three phenological stages of the crop.

	50% Covered floor	Floral button 1 mm	Pre-harvest
Biomass (Mg/ha)	1	2	4
NBI (Ud. Dualex)	34-36	31-33	27-29

The values of the IBW for each trial of the Barcelona and Typical varieties are shown below, as a function of crop biomass, with respect to the yield curve developed (Figures 90, 91, 92, 93 and 94).

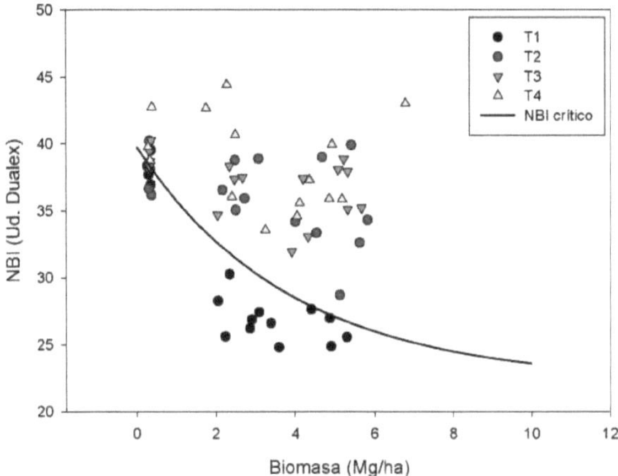

Figura 90.   Analytical curve and NBI index of Dualex as a function of crop biomass for the trial of the Barcelona variety in 2012.

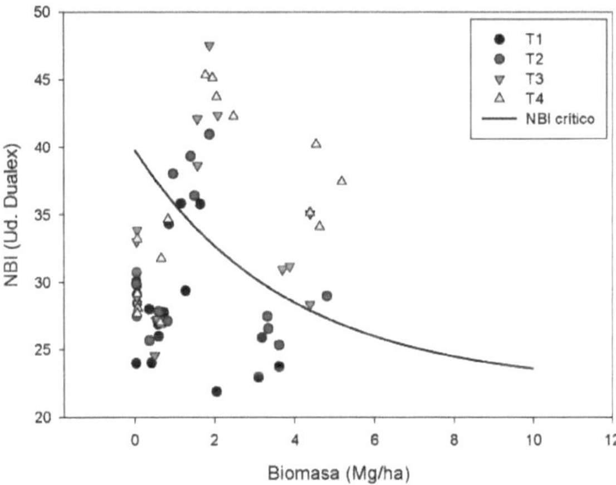

Figura 91.   Analytical curve and NBI index of Dualex as a function of crop biomass for the Barcelona variety trial in 2013.

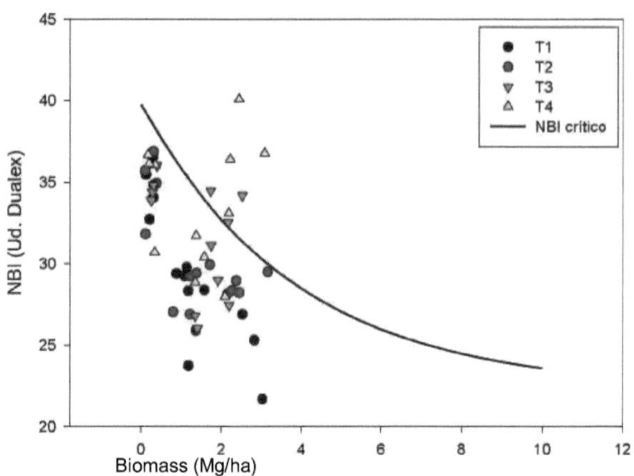

Figura 92.   Analytical curve and NBI index of Dualex as a function of crop biomass for the trial of the Barcelona variety in 2014.

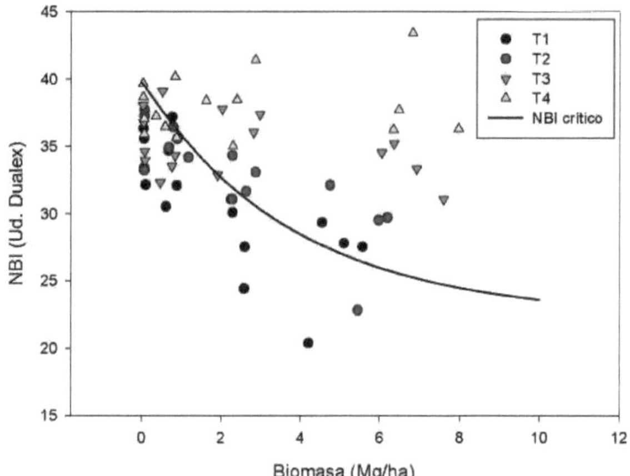

Figura 93.   Analytical curve and NBI index of Dualex as a function of crop biomass for the Typical variety trial in 2013.

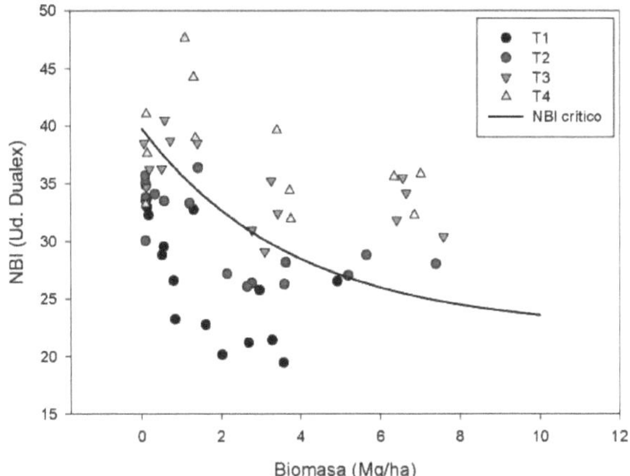

Cntic curve and NBI index of Dualex as a function of crop biomass for the Typical variety trial in 2014.

In each trial, the methodology used in grapevine by Cerovic *et al.* (2015) was used, in which the model was evaluated according to the parameters of sensitivity, specificity, Youden's index and precision index. The sensitivity of the model is defined as the percentage of treatments with higher values of NBI and higher values of yield, correctly identified. The specificity refers to the percentage of treatments with lower BIL values and lower production values correctly identified. On the other hand, the Youden (Youden, 1950) and precision (Cerovic *et al.*, 2015) indices relate the sensitivity and specificity of the model to classify it according to the suitability of the adjustment made. For index values close to 1 the fit will be more robust, and the opposite for values close to 0. Table 45 shows the values of sensitivity, specificity, Youden index and precision index for all the tests performed.

Table 45. Dualex NBI curve. Value of the sensitivity, specificity, Youden and precision indexes in the assays performed during 2012, 2013 and 2014 with the Barcelona and Typical varieties.

	Var. Barcelona			Var. Typical	
Mdices	2012	2013	2014	2013	2014
Sensitivity	1,00	1,00	0,44	1,00	1,00
Specificity	0,92	0,69	0,95	0,44	0,74
Youden Index	0,92	0,69	0,39	0,44	0,74
Precision Index	0,98	0,84	0,72	0,72	0,86

The specificity of the model, which classifies the treatments with lower yields and lower BIN values, allows us to assess the suitability of the model to identify the deficit treatments that could be corrected by fertilization. Therefore, of all the values in the table, we will focus mainly on the specificity of the model for the possible detection of deficiencies.

Table 45 shows that for the Barcelona variety in 2012 and 2014 very high specificity values were obtained, 0.92 and 0.95. However, for the year 2013 in this variety the specificity value was significantly lower, which may be due to the effect of hail on the plants, causing moderate defoliation and altering the vegetative cycle of cauliflowers.

In the Typical variety, the behavior is similar to that of the Barcelona variety, showing very high specificity values in 2014. In 2013, possibly due to hail, the values dropped significantly.

The Youden indices and model precision show high values for the adjustments made in each trial. Only in 2014 for the Barcelona variety and in 2013 for the Typical variety were values for the Youden index significantly lower than in the rest of the trials. This is due to the fact that in 2014, for the Barcelona variety, the sensitivity of the model obtained lower values than the rest of the trials, and in 2013, for the Typical variety, the specificity was significantly lower than in the rest of the experiments, possibly due to the effect of hail.

The accuracy index has shown very high values for the model in all tests performed.

5.8. MULTIPLEX sensor

The results obtained with the MULTIPLEX sensor have been similar to those obtained with the DUALEX sensor in detecting differences between treatments before top dressing fertilization and in the twenty days after fertilizer application, which could be useful to correct possible nitrogen fertilization deficiencies depending on the length of the crop cycle. The MULTIPLEX sensor, based on fluorescence ratios, provides two measures of chlorophyll content SFR and nitrogen balance NBI. Like DUALEX, the NBI index seems to be more discriminating than the SFR index, but the use of both allows a better diagnosis of plant status in relation to nitrogen fertilization.

Sampling with MULTIPLEX is fast as it is a fluorescence measurement method and although a manual detector has been used in these tests, it is possible to install it in a vehicle allowing for walk-throughs.

In the Barcelona variety (Table 46) the results are similar to those observed in the case of Dualex. The initial SFR index was around 5 units and was maintained in the fertilized treatments, while it decreased by about 5% in the unfertilized treatments. The NBI index started from values close to 1.6 units and has maintained this value throughout the crop in the fertilized treatments while it has decreased up to 33% on average in the unfertilized treatments. As with Dualex, there was a gradation between T1 and T4 and also the NBI index was more discriminating.

Table 46. Mean initial and final Multiplex values of the fertilized (F) and unfertilized (NF) treatments in the Barcelona variety trials in 2012, 2013 and 2014, with the percentage of variation (±A).

	Initial SFR	Final SFR	±A	Initial NBI	Final NBI	±A
Multiplex F	5	5	=	1.6	1.6	=
Multiplex NF	4.8	4,5	-4.8%	1.4	1.0	-32.8%

The SFR value of Multiplex for the Typical variety (Table 47) remained stable in the fertilized treatments and decreased by 16.4% in the unfertilized treatments. The NBI index of this sensor increased to 16.6% in the fertilized treatments and decreased sharply in the unfertilized treatments to 40.8%.

Table 47. Mean initial and final Multiplex values of the fertilized (F) and unfertilized (NF) treatments in the 2013 and 2014 Typical variety trials, with the percentage of variation (±A).

	Initial SFR	Final SFR	±A	Initial NBI	Final NBI	±A
Multiplex F	4.7	4.7	=	1.4	1.7	16.6%
Multiplex NF	5.0	4.2	-16.4%	1.3	0.8	-40.8%

As in Dualex, in the two varieties, chlorophyll content is maintained or slightly increased in the fertilized treatments and decreased in the unfertilized ones. The NBI index is shown to be more discriminating due to the introduction of the stress factor provided by the flavonoid index used in the calculation. This trend of the NBI index coincides with that observed in the trial of Galambosova *et al.* (2014) in wheat.

5.8.1. Multiplex critical curve

As with Dualex, after analyzing the results obtained in trials with the Multiplex sensor, which also showed sensitivity to significantly differentiate the deficit treatments, it was observed that there was a significant linear correlation between the values of the Multiplex NBI index and the concentration of total nitrogen in the leaf throughout the development of the crop on the different measurement dates. Galambosova *et al.* (2014) found high linear correlations between Multiplex NBI index and plant leaf nitrogen concentration in wheat.

Figure 95 compares the values obtained for the Multiplex NBI index and total leaf nitrogen concentration for the Barcelona variety during 2012, 2013 and 2014. Figure 96 compares the values for the Typical variety during 2013 and 2014.

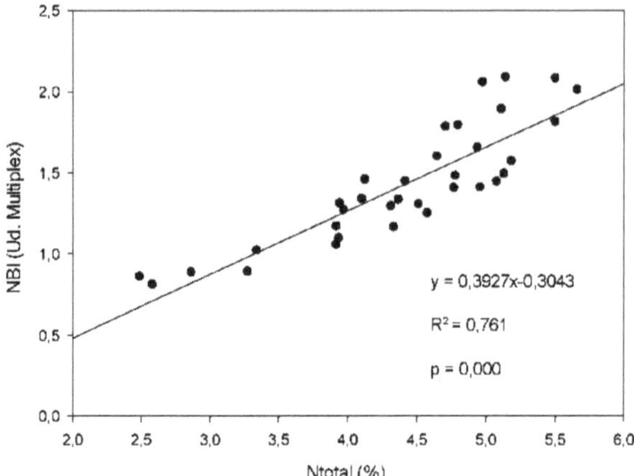

Figure 95. Relationship between Multiplex NBI index and total nitrogen concentration in cauliflower leaf of the Barcelona variety during the years 2012, 2013 and 2014. The p-value indicates the degree of statistical significance.

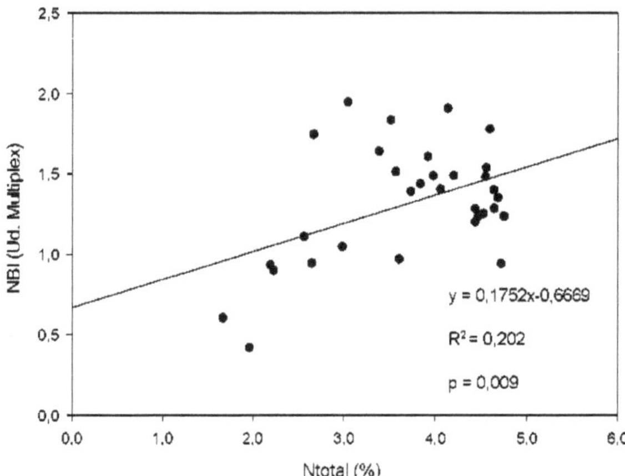

Figura 96. Relationship between Multiplex NBI index and total nitrogen concentration in cauliflower leaves of the Typical variety during 2013 and 2014. The p-value indicates the degree of statistical significance.

A comparison was made of the regression parameters for both varieties and it was found that there were no significant differences between them. Therefore, it was possible to establish a common regression for both varieties, the result of which is shown in Figure 97:

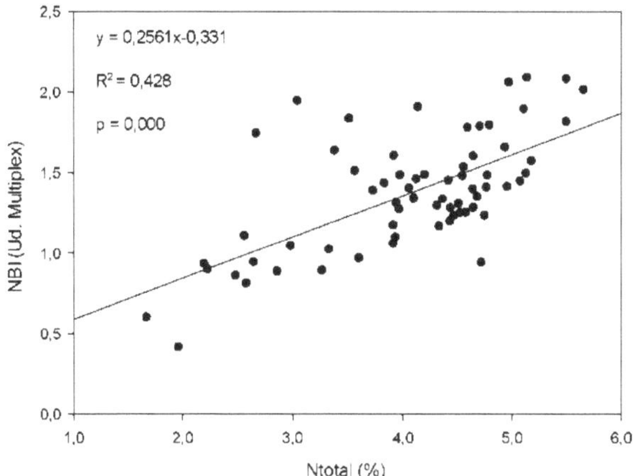

Figura 97. Relationship between Multiplex NBI index and total nitrogen concentration in cauliflower leaves of Barcelona and Typical varieties. The p-value indicates the degree of statistical significance.

Once the existence of a linear relationship, very significant, between both parameters has been verified, a function of NBI has been adjusted according to the biomass of the crop, following the same methodology used with the Dualex sensor described in the previous section. The function obtained is:

$$y = 0.973 \left(1 + e\text{-}0.4^{67x}\right)[10]$$

where "y" is the critical value of the NBI index, "x" *is* the biomass of the crop.

regression and parameters being highly significant (Table 48).

Sum of squares (SC), degrees of freedom (gdl), mean squares (MS) and value of the F statistic. Value of parameters "a" and "b" of the function [10], standard error (SE) and value of the t statistic (tobserved). Significance: *** = p<0.001.

Source	SC	gdl	CM	F	Parameter		SE	tobservada
Regression	38,164	2	19,082	596,3***				
Error	0,55	17	0,032		a =	0,973	0,072	13,492***
Total	38,713	19			b =	0,467	0,178	2,618*
R² corrected	0,747							

Figure 98 shows the common model obtained for the two varieties, as well as the upper and lower limits of the confidence interval (95%) of the model.

As in Dualex, starting from a biomass of approximately 1 Mg/ha, the curve identifies 92% of the deficit treatments (Figure 98). Thus, from a given value of crop biomass, this curve can determine the threshold values below which the cauliflower crop is in a state of nitrogen nutritional deficit.

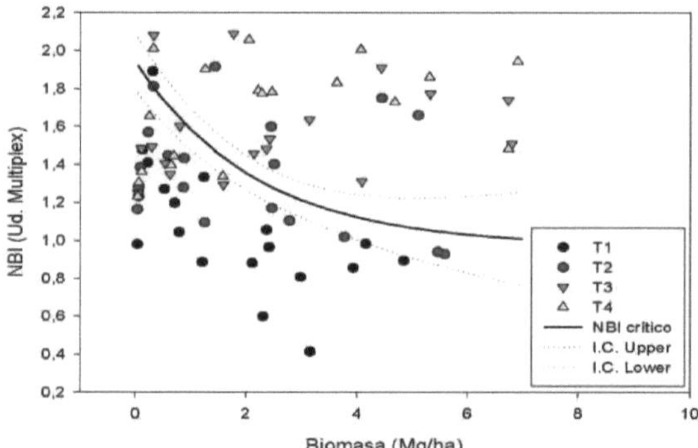

Figura 98. Multiplex IBN index and biomass (Mg/ha) of the Barcelona and Typical varieties in the 2012, 2013 and 2014 trials. Adjusted BKN curve as a function of crop biomass (cntic BKN) and 95% confidence interval (C.I. Upper and C.I. Lower).

Taking into account the values of the critical curve obtained, the confidence interval of the model and the biomass of the crop, we can propose minimum sufficiency values (Table 49).

Table 49. Minimum sufficiency values for the Multiplex NBI index as a function of biomass, at three phenological stages of the crop.

	50% Covered floor	Floral button 1 mm	Pre-harvest
Biomass (Mg/ha)	1	2	4
NBI (You Multiplex)	1,4-1,6	1,2-1,4	0,9-1,0

The values of the IBW for each trial of the Barcelona and Typical varieties are shown below, in terms of crop biomass, with respect to the critical curve (Figures 99, 100, 101, 102 and 103).

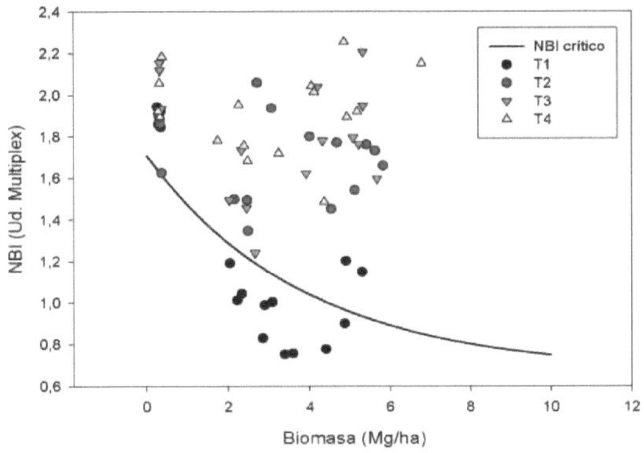

Figura 99. Multiplex NBI index curve as a function of crop biomass for the Barcelona variety trial in 2012.

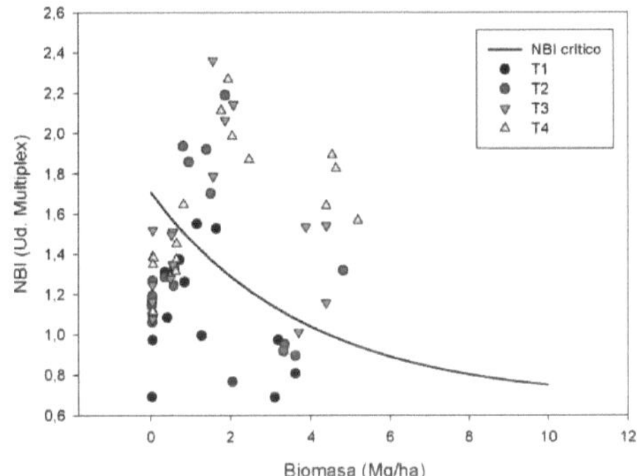

Figura 100. Multiplex NBI index curve as a function of crop biomass for the Barcelona variety trial in 2013.

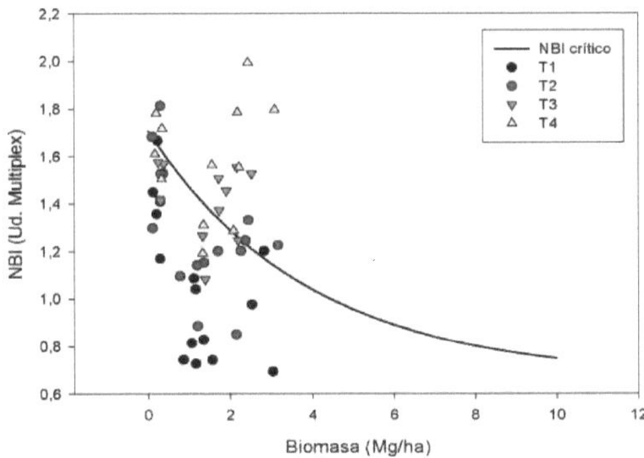

Figura 101. Multiplex NBI index curve as a function of crop biomass for the Barcelona variety trial in 2014.

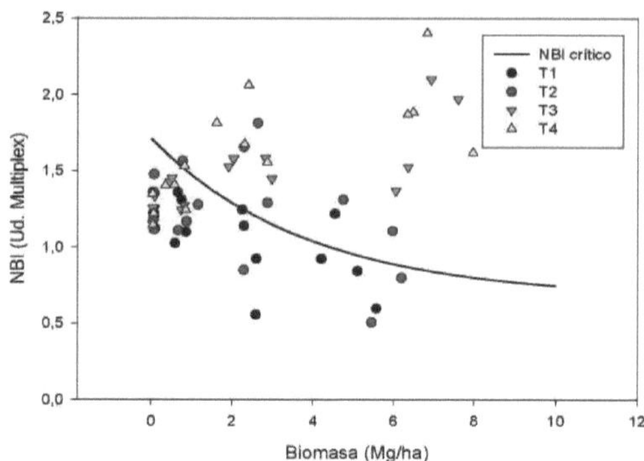

Figura 102. Multiplex NBI index curve as a function of crop biomass for the Typical variety trial in 2013.

Figura 103.

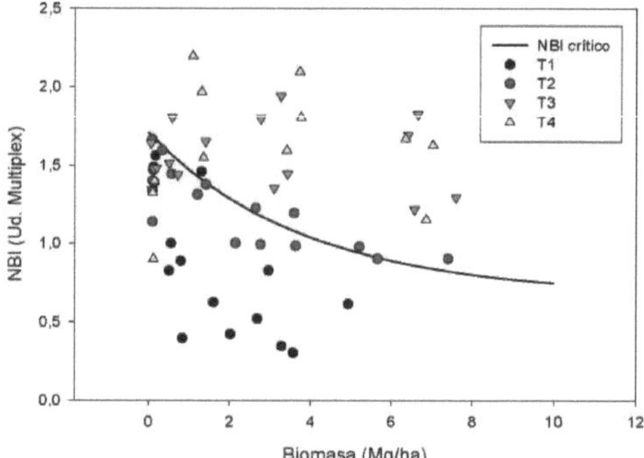

Cntic curve and Multiplex NBI index as a function of crop biomass for the Typical variety trial in 2014.

In each test performed, the methodology used for Dualex in section 5.7 has been used, in which the model is evaluated according to the parameters of sensitivity, specificity, Youden index and precision index. Table 50 shows these values for all the tests performed.

Table 50. Multiplex NBI diagnostic curve. Value of the sensitivity, specificity, Youden and precision indexes in the assays performed during 2012, 2013 and 2014 with the Barcelona and Typical varieties.

	Var. Barcelona			Var. Typical	
Mdices	2012	2013	2014	2013	2014
Sensitivity	0,97	0,94	0,56	1,00	1,00
Specificity	0,83	0,56	0,90	0,63	0,95
Youden Index	0,81	0,50	0,46	0,63	0,95
Precision Index	0,94	0,75	0,75	0,81	0,97

The specificity of the model, which classifies the treatments with lower yields and lower

BIL values, allows us to assess the quality of the model in order to identify the deficit treatments that could be corrected by fertilization. Therefore, of all the values in the table, for the possible detection of deficiencies, we will focus mainly on the specificity of the model.

Table 50 shows that for the Barcelona variety in 2012 and 2014 very high specificity values were obtained (0.83 and 0.90). However, for the year 2013 in this variety the specificity value was significantly lower, which may be due to the effect of hail on the plants, causing moderate defoliation and altering the vegetative cycle of cauliflowers.

In the Typical variety, the behavior is similar to that of the Barcelona variety, showing very high specificity values in 2014. In 2013, possibly due to hail, the values dropped significantly.

The Youden and precision indices of the model show high values for the adjustments made in each trial. Only in 2014 in the Barcelona variety were values for the Youden index lower than 0.5 obtained. This is due to the fact that in 2014, for the Barcelona variety, the sensitivity of the model obtained lower values than the rest of the trials.

The accuracy index has shown very high values for the model in all tests performed.

5.9. CROP CIRCLE sensor

The CROP CIRCLE sensor, based on reflectance ratios, provides two indices for chlorophyll content and crop biomass, NDRE and NDVI. When there are differences in growth, the NDVI index is a good detector and at more advanced stages of the crop, the NDRE index detects color differences caused by different leaf chlorophyll contents. By providing reflectances in the red, far-red and near-infrared bands, it allows the calculation of other indices, such as REDVI, which have high correlations with the nutritional status of the plant.

The NDVI index (Rouse *et al.*, 1973) is a vegetation index used to estimate plant biomass. Figure 104, with data from the initial and final sampling dates of the 2013 and 2014 trials on the Barcelona and Typical varieties, shows the relationship between the biomass of the Barcelona and Typical varieties and the NDVI index, which shows a good linear correlation.

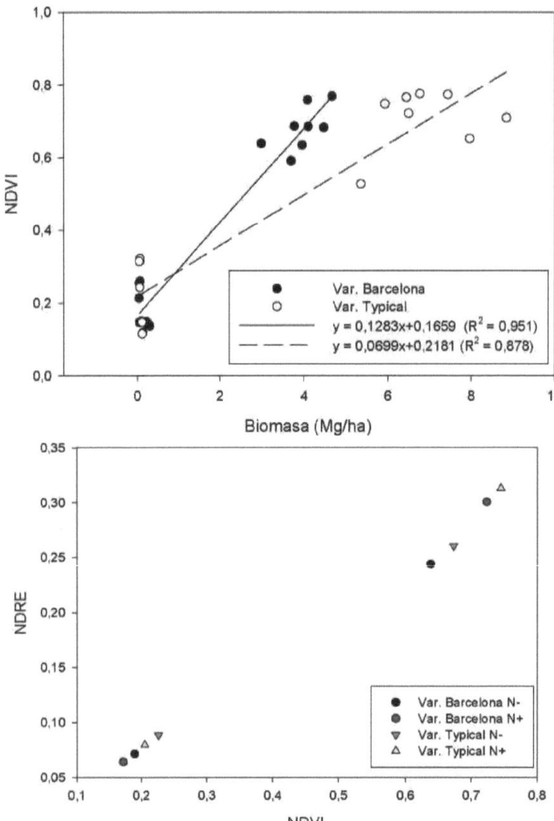

Figura 104.   Relationship between NDVI index and crop biomass in Mg/ha for Barcelona and Typical varieties.

The NDRE index (Barnes *et al.*, 2000) is an estimator of chlorophyll content and indirectly of nitrogen content. The relationship between this index and the NDVI index for the most fertilized (mean of T3 and T4) and least fertilized (T1 and T2) treatments in the years 2013 and 2014 are shown in the following figure:

Figura 105.   Relationship between NDVI index and NDRE index for Barcelona and Typical varieties in 2013 and 2014. N+: most fertilized treatments, T3 and T4; N-: least fertilized treatments.

In all treatments, in both varieties there is an increase in NDVI and NDRE values in parallel to crop development. This increase is due to the increase in reflectance in the near infrared (NIR) region along with the decrease in reflectance in the red (R) and far red (F) region. Plants absorb photons of red and blue light to carry out their photosynthetic activity, so the reflectance in the red will be higher in stressed plants and lower in healthy plants (due to the absorption of photons of that wavelength by chlorophylls). In the near infrared the trend is the opposite, healthy plants have a higher reflectance and stressed plants absorb a greater amount of this radiation. Casa *et al.* (2014), found high correlations between leaf chlorophyll content and the ratio of NIR and Red-Edge reflectances in wheat, beans and ma^z.

The initial reflectance in the NIR of 27% in both varieties increases to values of 37 and 39% in the Barcelona and Typical varieties, in the most fertilized treatments, and 34 and 28% in these varieties in the least fertilized treatments. The initial reflectance in the R of 19% in both varieties decreases to around 5.5%, also in both, in the fertilized treatments and with less intensity in the less fertilized ones, to around 7%. In the RE region, the initial reflectance values are similar in both varieties, around 23%, and decrease to values of 19% in the most fertilized treatments and 21% in the least fertilized. This evolution of reflectance values in the R, RE and NIR regions causes a greater increase in NDVI and NDRE values in the most fertilized treatments in relation to the least fertilized ones.

The results obtained with the CROP CIRCLE sensor have shown to be more discriminant than those obtained with the SPAD, DUALEX and MULTIPLEX sensors for detecting differences between treatments during the whole period of measurements, from before the mulch fertilization until the beginning of the harvest, since it has been able to detect differences between treatments before the rest of the equipment and due to the number of samples taken, carried out continuously, it presents higher statistical significance values than the rest of the equipment and a lower sampling error.

The sampling with CROP CIRCLE is fast as it is a method of measurement by reflectance, being able to be installed on vehicles and allowing to make tours and a wide sampling of the plots.

Crop Circle also showed sensitivity to significantly differentiate the deficit treatments, but it was also observed that there was a significant correlation between the values of the REDVI index (Tucker *et al.*, 1979) and the nitrogen nutrition index NNI, (Lemaire and Gastal, 1997). This correlation was also found by Cao *et al.* (2013) in a rice trial. The REDVI index is expressed as the difference in reflectance between the NIR (near infrared) and the Red-Edge. This index, as well as many others such as NDRE (Barnes *et al.*, 2000), RESAVI (Sripada *et al.*, 2006) or RERDVI (Roujean *et al.*, 1995) focus on the region of the spectrum between the red-edge and the near-infrared, where, as mentioned in the introduction of this work, greater differences in light absorption are found between plants.

The NNI index is calculated as the quotient between the total nitrogen concentration and the total nitrogen concentration, below which the plant is in nutritional deficit. When this index presents values below 1, the plant is in nutritional deficit and above this value we find ourselves in conditions of luxury consumption by the plant. This applies for biomasses greater than 1 Mg/ha, from which the concentration of cntic nitrogen depends on the aerial biomass according to studies cited by Justes *et al.* It was observed that, for the Barcelona variety during the years 2012, 2013 and 2014, for the Typical variety in the years 2013 and 2014, and jointly for the two varieties, there was a highly significant correlation between the values of the REDVI and NNI indices (Figures 106, 107). This coincides with the results found by Cao *et al.* (2013) in this region of the spectrum, throughout the cycle of a rice crop.

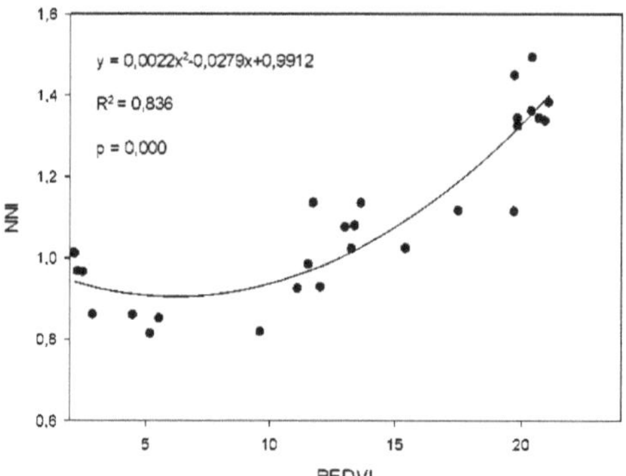

Relationship between the NNI and REDVI indices for the Barcelona variety during 2012, 2013 and 2014. The p-value indicates the degree of statistical significance.

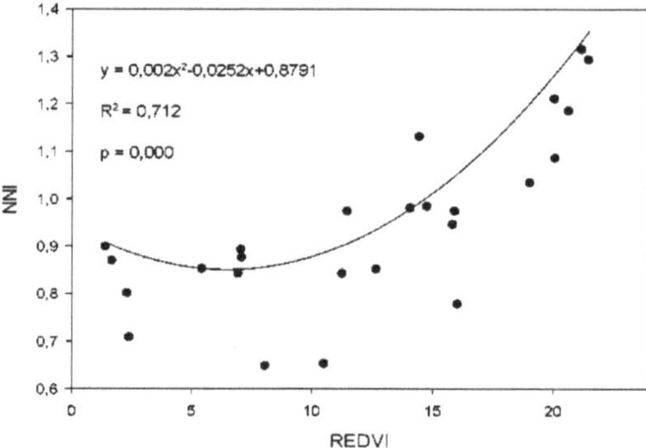

Figura 107. Relationship between the NNI and REDVI indices for the Typical variety for the years 2013 and 2014. The p-value indicates the degree of statistical significance.

A comparison of the regression parameters for both varieties was made and it was found that there were no significant differences between them. Therefore, it was possible to establish a common regression for both varieties, the result of which is shown in Figure 108.

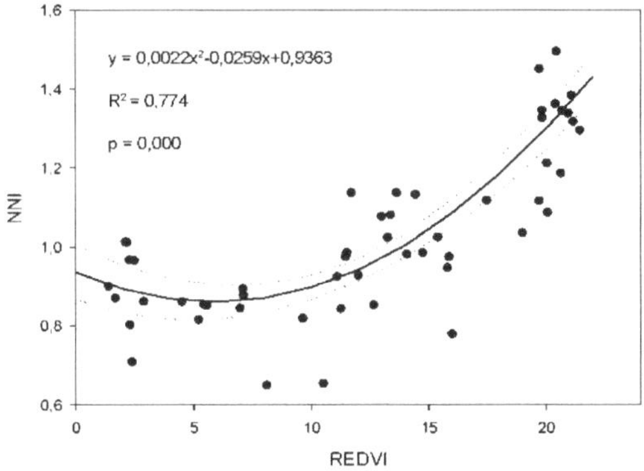

Figura 108. Relationship between NNI and REDVI indices for Barcelona and Typical varieties. The p-value indicates the degree of statistical significance. The dotted curves parallel to the regression represent the 95% confidence interval.

The figures show that from the moment when approximately 10 units of the REDVI index are reached, the slope of the function that relates the NNI and REDVI indexes changes significantly. This is due to the fact that up to that moment the plant has not reached a biomass higher than 1 Mg/ha, coinciding with 50% soil cover. For lower biomasses (in the early stages of vegetative development of the plant) the concentration of cntic nitrogen is independent of aerial biomass according to studies cited by Justes et al. (1994), and the NNI index is directly related to the concentration of cntic nitrogen. From that biomass of 1 Mg/ha (50% of soil covered) the REDVI index can be a good estimator of the NNI index, so it could be used together with the NDVI and NDRE indexes to differentiate treatments with a deficient nutritional status.

5.10. Nitrate in sap

Before top dressing, the mean N-NO3 concentration⁻ in sap in all treatments was 2100 ± 873 ppm. It normally ranged between 1300 and 1500 ppm except in two trials where very high values of 2500 and 5500 ppm were reached. In general terms, this concentration decreased throughout the crop until harvest, depending on the available nitrogen. Treatment T1, unfertilized, ended the crop with values close to zero, while the most fertilized crops showed higher values.

The concentration of N-NO3⁻ in sap was a very sensitive and repeatable indicator, capable of showing significant differences between treatments, especially after top dressing fertilization. This determination was carried out approximately one month after transplanting. Nitrate concentration before top dressing fertilization did not show significant differences in any of the trials and therefore did not help to adjust this fertilization. Like other in-plant measurements, it can be affected by other factors e.g. water deficit, etc. It is a destructive measurement that requires time and facilities to perform. It has been described as a valuable and rapid technique for estimating nitrogen needs (Kubota *et al.*, 1997), other authors (Olsen and Lyons, 1994) concluded that sap nitrate concentration is a sensitive indicator of plant nitrogen nutritional status as well as variations in soil nitrate content. It has also been criticized for the high variability of its results and the lack of agreement of cntical levels (Westerveld *et al.*, 2003).

5.10.1. N-NO3 curve in sap

For the construction of the sap N-NOa" content curve over the course of the crop, the relative results of both crop time and N-NO3 concentration‾ were used. These results were associated with total pellet yield. Significantly less productive treatments were discarded and from the most productive treatments the results with the lowest values were chosen, following the methodology indicated by Olasolo (2013) to adjust the nitrogen curve in jud^a. With these data, a non-linear model of the type y = B EXP (-C x) was fitted, where y is the relative N-NO3 content‾ (%) and x is the relative cultivation time (%). A common model was fitted for the shorter-cycle varieties, Barcelona and Casper, and another for the medium-cycle cv. Typical. For the short-cycle varieties, the adjusted function was as follows:

$$y = 182.7\ e^{-0,032\,x} \qquad\qquad [11]$$

The regression and parameters are highly significant (Table 51 and Figure 109 A).

Table 51. Sum of squares (SC), degrees of freedom (gdl), mean squares (MS) and value of the F statistic. Value of the parameters "a" and "b" of the function [11], standard error (SE) and value of the t statistic ($t_{observed}$). Significance: *** = $p<0.001$.

Source	SC	gdl	CM	F	Parameter		SE	tobservada
Regression	89.342,9	2	44.671,5	316,4***				
Error	4.800,1	34	141,2		a =	182,771	21,003	8,70***
Total	94.143,0	36			b =	0,032	0,003	9,83***
R^2 corrected	0,828							

From the adjusted model, the critical values for the Barcelona and Casper varieties will be reached below 1195, 651, 488 and 292 ppm N-NOa" at 10%, 50% soil cover, 1 mm flower bud and the beginning of harvest, respectively. In the case of short-cycle varieties, the values found are similar to those recommended by the model of Kubota *et al.* (1997). Prior to the first harvest, cv. Barcelona and Casper present a cntic value of 292 ppm, coinciding with the values described by Hochmuth in 2015 (300-500 ppm) and Kubota *et al.* in 1997 (290 ppm). Following the methodology described above, in the Dualex and Multiplex sensors, for the calculation of sensitivity, specificity and Youden and precision indices, the model for the Barcelona and

Casper varieties discriminates 77% of the non-deficient treatments and 70% of the deficient ones (Table 52).

Table 52. Nitrate in sap test curve. Value of the sensitivity, specificity, Youden and precision indexes in the tests carried out with the Barcelona and Casper varieties.

Sensitivity	0,77
Specificity	0,70
Youden Index	0,46
Precision Index	0,73

For the medium cycle Typical variety, the function was adjusted taking into account the lack of experimental results beyond the 1 mm flower bud stage. The adjusted function was as follows:

$$y = 120.9\ e\text{-}0.0^{41\,x} \qquad [12]$$

with the regression and parameters being highly significant (Table 53 and Figure 109 B).

Sum of squares (SC), degrees of freedom (df), mean squares (MS) and value of the F-statistic. Value of the parameters "a" and "b" of the function [12], standard error (SE) and value of the t-statistic (tobserved). Significance: *** = p<0.001.

Source	SC	gdl	CM	F	Parameter		SE	tobservada
Regression	87.570,7	2	43.785,3	182,7***				
Error	6.232,3	26	239,7		a =	120,875	13,252	9,12***
Total	93.803,0	28			b =	0,041	0,006	6,56***
R^2 corrected	0,676							

The parameters that evaluate the suitability of the model (Table 54) indicate that for this variety a higher sensitivity has been obtained with respect to the model obtained for short-cycle varieties, but a lower specificity.

Table 54. Nitrate in sap test curve. Value of the sensitivity, specificity, Youden and precision indexes in the tests carried out with the Typical variety.

Sensitivity	0,89
Specificity	0,35
Youden Index	0,24
Precision Index	0,56

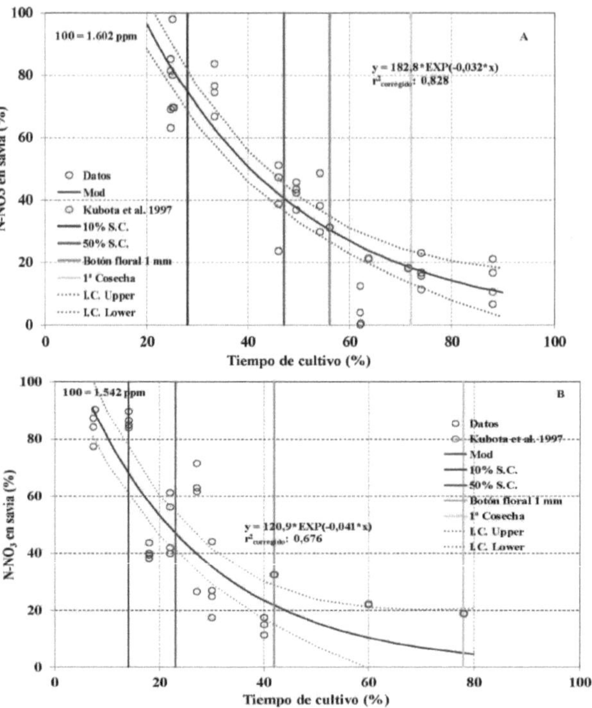

Figura 109. Relative cnitic N-NO3 content in sap (%) as a function of relative cultivation time (%). A: cv. Barcelona and Casper. B: cv. Typical. S.C.: percentage of soil covered. The recommendations of Kubota *et al.* (1997), from results of Doerge *et al.* (1991), are included.

The nitrate concentration in the sap of each trial of the Barcelona, Casper and Typical varieties is shown below, as a function of the percentage of the crop, with respect to the obtained yield curve (Figures 110, 111, 112, 113, 114 and 115).

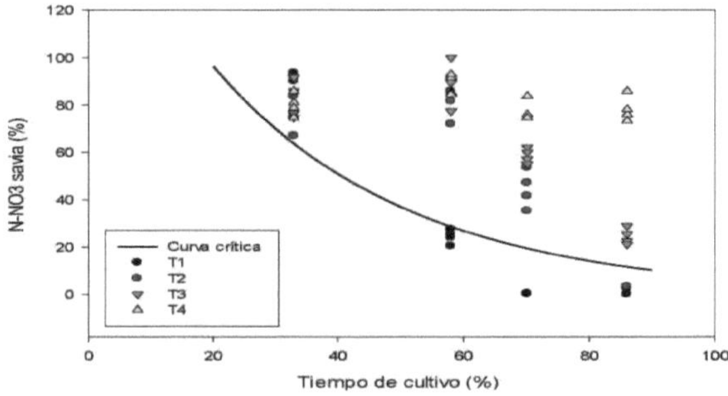

Figura 110. Critical curve and nitrate content in sap as a function of cultivation time for the Barcelona variety trial in 2012.

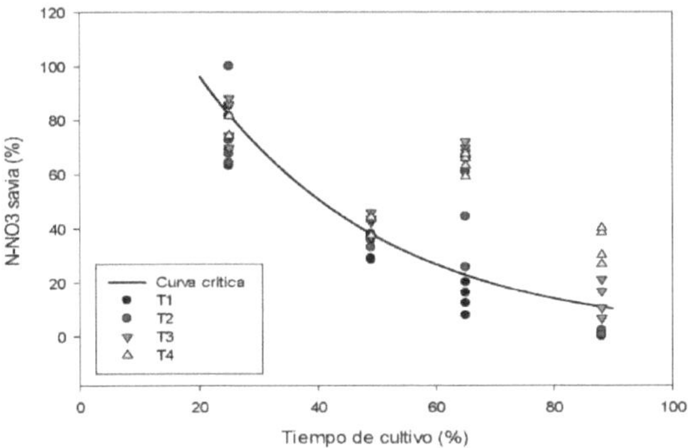

Figura 111. Critical curve and nitrate content in sap as a function of cultivation time for the Barcelona variety trial in 2013.

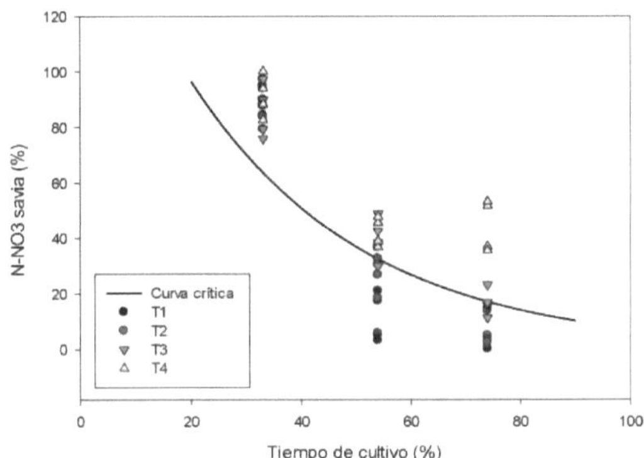

Figura 112. Cntic curve and nitrate content in sap as a function of cultivation time for the Barcelona variety trial in 2014.

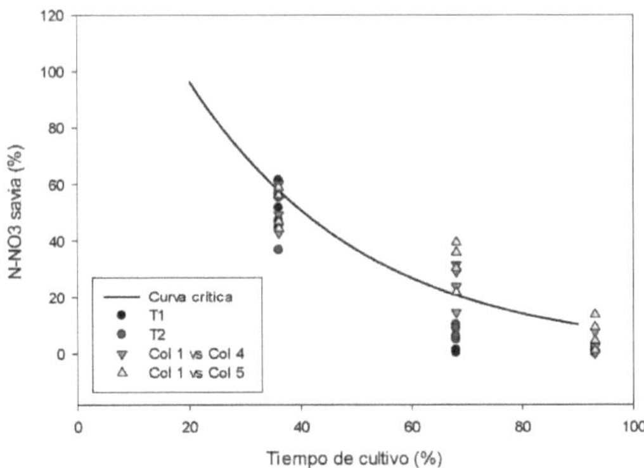

Figura 113. Cntic curve and nitrate content in sap as a function of cultivation time for the Casper

variety trial in 2014.

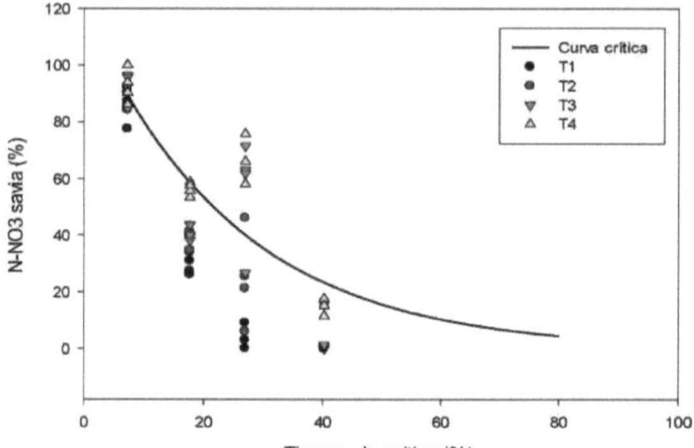

Cntic curve and nitrate content in sap as a function of cultivation time for the Typical variety trial in 2013.

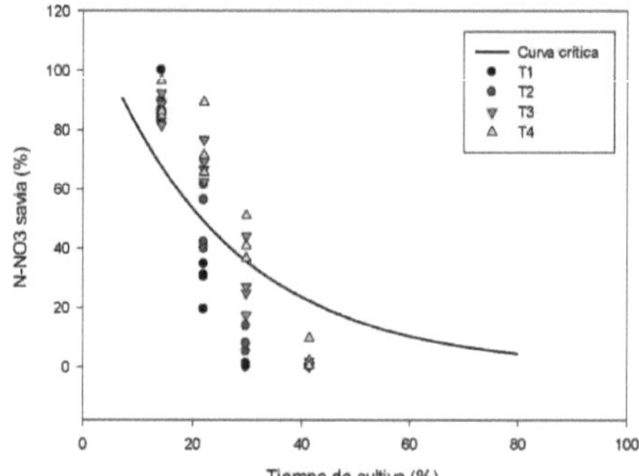

Cntic curve and nitrate content in sap as a function of cultivation time for the Typical variety trial in 2014.

To check the suitability of the model in each test performed, the model is evaluated according to the parameters of sensitivity, specificity, Youden index and precision index. Table 55 shows the values of sensitivity, specificity, Youden index and precision index for all the tests performed.

Table 55. Nitrate in sap test curve. Value of the sensitivity, specificity, Youden and precision indexes in the tests carried out for all the assays.

	Barcelona			Casper	Typical	
	2012	2013	2014	2014	2013	2014
Sensitivity	0,92	0,72	0,83	0,46	0,86	0,89
Specificity	0,75	0,78	0,37	0,96	0,38	0,35
Youden Index	0,67	0,50	0,20	0,42	0,23	0,24

| Precision Index | 0,88 | 0,75 | 0,57 | 0,71 | 0,60 | 0,56 |

Table 55 shows that very high specificity values (0.75 and 0.78) were obtained for the Barcelona variety in 2012 and 2013. However, for the year 2014 for this variety, the specificity value was significantly lower. In the Casper variety, in 2014, very high specificity values were also obtained (0.96). The sensitivity of the model was high in all trials, except for the Casper variety. And the precision index of the model has shown values higher than 50 % in all trials.

Final considerations

For some indices of the DUALEX, MULTIPLEX and CROP CIRCLE equipment, high correlations have been found with the nitrogen content of the plant, which has been used to adjust models that relate the measurement to the biomass of the crop, depending on the phenological stage of the crop. The critical curves obtained for the sensors show that from a biomass of approximately 1 Mg/ha (50% of the soil covered) the measurement stabilizes and it is from this moment when deficit treatments begin to be distinguished from non-deficit treatments, as occurs with the total nitrogen content in the leaf. Therefore, these models could be used to determine a nitrogen nutritional deficit and to be able to correct it by fertilization. The methodology, described in the previous sections, could be implemented in other varieties and/or crops, with a greater number of trials and a greater number of samples, to determine critical values and to use this equipment as estimators of the nitrogen content in the plant.

These deficiencies should be detected in time to be corrected in blanket fertilization. The time of application of this correction will depend on the crop cycle, its stage of development before the beginning of pellicle formation, and the fertilizer application technique, and may be extended in time in case of fertigation.

As for the destructive measures, the concentration of total nitrogen in cauliflower leaves and $N\text{-}NO_3^-$ in sap showed similar behavior in all trials. In spite of being destructive and laborious methods, they were able to discriminate the deficit treatments throughout the crop cycle in the different trials.

Of all the methods used in the study, the results obtained for the Nmin analysis confirm the usefulness of the method for the design of nitrogen fertilization of cauliflower, as well as the importance that mineralized nitrogen can acquire in the balance, the possible losses by volatilization and the need to reduce losses by leaching through correct irrigation scheduling. This is the method that offers the most information of those studied for decision making in a correct nitrogen fertilization, since it is the only one that allows us to know the nitrogen content available in the soil.

The results observed with the nondestructive measurement equipment complement the Nmin results and demonstrate the usefulness of these methods for detecting deficiencies in the nutritional status of plants. Of these, the CROP CIRCLE ACS-430 reflectance meter has allowed continuous analysis of a large number of samples in a short period of time, thus reducing the sampling error and obtaining more representative values of the crop than with the other equipment used. This equipment also allows taking samples from a vehicle and georeferencing these measurements, thus obtaining precise information of our plot in order to program a differentiated fertilization.

Taking into account all the results obtained, the only method capable of quantitatively specifying a fertilizer recommendation today is the Nmin analysis. Delaying the soil analysis until moments before the top dressing fertilization, allows us to adjust the nitrogen fertilization more efficiently, since we take into account the possible mineralization of the soil organic matter from the beginning of the crop, as well as the possible losses by leaching.

In my opinion, in the next few years, and due to the exponential development of technology, the use of reflectance equipment, both at laboratory and field level, capable of making fast and precise measurements along the electromagnetic spectrum will represent a qualitative leap in the knowledge of spectral radiation and in the analysis of the information that plants reflect, absorb or transmit in the form of light.

It is necessary a basic work in the study of information of specific wavelengths that help us to design equipment with a greater discriminating capacity on the physiological processes occurring in plants and to apply this technology developed from the laboratory to the field and then implement it in more versatile systems such as drones, unmanned aircraft or satellite imagery.

The objective of these teams should focus not only on the early detection of nutritional deficiencies, but also on the quantification of these deficiencies in order to correct them quantitatively.

CONCLUSIONS

Conclusions

1.   In the Barcelona variety, the mean value of Navailable Nabove which it is not response in the production was 184 ± 20 kgN/ha. In the Typical variety in 2013, this amount was 189 ± 45 kgN/ha. In Casper in 2012, no differences in production were obtained and in 2014, for this same variety, the value found was 143 ± 7 kgN/ha, these values were significantly lower than in the rest of the trials.

2.   The average concentration of total nitrogen in cauliflower leaves of the variety Barcelona was above 4.5% at the beginning of the crop and around 3.5% at harvest. In the Typical variety, these average values were between 4.5% and 3%. In the Casper variety the average values were between 5.3% and 4%, and 2.5% in two different years. This concentration has shown a similar behavior in all the trials and has been able to discriminate the deficit treatments throughout the crop cycle in the different trials.

3.   The relationship between total nitrogen and crop biomass has been able to discriminate, through the nitrogen curve, the treatments considered as deficient. This discrimination was more evident from biomasses higher than 1 Mg/ha.

4.   The average extractions of the cauliflower varieties studied were as follows 246 kg of nitrogen per hectare.

5.   In the Barcelona variety trials, the mineralization of the material The soil organic matter measured in the field reached an average value of 46 kg N/ha for the crop in the first 30 cm of the soil. Nitrogen supply through the mineralization of soil organic matter can account for about 20% of the plant's nitrogen extraction.

6.   The results of the balance sheet analysis in the different tests confirm the usefulness of the Nmin method for the design of nitrogen fertilization of cauliflower, as well as the importance that mineralized nitrogen can acquire in the balance, the possible losses by volatilization and the correct irrigation scheduling to reduce losses by leaching.

7.   Delay soil testing until just prior to fertilization of the soil. The use of a mulch cover allows a more efficient adjustment of the nitrogen fertilization, since we take into account the possible mineralization of the soil organic matter from the beginning of the crop, as well as the possible losses due to leaching.

8.   The results obtained with the SPAD sensor have been highly variable and of low repeatability.

9.   The results obtained indicate that the measurements with the DUALEX sensor have The results were more repeatable and sensitive than those obtained with the SPAD sensor for detecting differences between treatments prior to top dressing and within 20 days of fertilizer application.

10.   A significant linear correlation has been observed between DUALEX NBI index values and total nitrogen concentration in the leaf during crop development.

11.   The proposed DUALEX NBI index curve correctly identifies 92% of non-nitrogen deficient treatments and 79% of nitrogen deficient treatments.

12.   The results obtained with the MULTIPLEX sensor were similar to those obtained with the DUALEX sensor for detecting differences between treatments.

13.   It was observed that there was a significant linear correlation between the values of the
MULTIPLEX NBI index and total leaf nitrogen concentration throughout crop development at the different measurement dates.

14.   The proposed MULTIPLEX NBI index curve identifies 92% of the deficient treatments.

15.   The results obtained with the CROP CIRCLE sensor have been shown to be more discriminant than those obtained with the SPAD, DUALEX and MULTIPLEX sensors in detecting differences between treatments during the entire measurement period, from before cover crop fertilization to the beginning of harvest.

16.   A significant correlation was observed between the values of the REDVI and NNI indices, from a biomass of more than 1 Mg/ha.

17.   The concentration of $N\text{-}NO_3^-$ in sap has been a very sensitive and repeatable indicator, capable of showing significant differences between treatments, especially after top dressing fertilization.

18.   The adjusted sap nitrate curve for short-cycle varieties discriminates 77% of the non-deficit treatments and 70% of the deficit treatments. In the case of the adjusted curve for the Typical variety, a higher sensitivity was obtained with respect to the model obtained for the short-cycle varieties, but a lower specificity.

BIBLIOGRAPHIC REFERENCES

Abuzeid, A.E., Wilcockson, S.J. (1989). Effects of sowing date, plant density and year on growth and yield of Brussels sprouts (*Brassica oleracea* var. *bull* at a subvar. *gemmifera*). The Journal of Agricultural Science, 112, 359-375.

Agati, G., Foschi, L., Grossi, N., Volterrani, M. (2015). In field noninvasive sensing of the nitrogen status in hybrid bermudagrass (*Cynodon dactylon x C. transvaalensis Burtt Davy*) by a fluorescence based method. Eur. J. Agron., 63, 89-96.

Agati, G., Foschi, L., Grossi, N., Guglielminetti, L., Cerovic, Z. G., Volterrani, M. (2013). Fluorescence-based versus reflectance proximal sensing of nitrogen content in *Paspalum vaginatum* and *Zoysia matrella* turfgrasses. Eur. J. Agron., 45, 39-51.

Allen, R.G., Pereira, L.S., Raes, D., Smith M. (1998). Crop evapotranspiration. Guidelines for computing crop water requirements. FAO Irrigation and drainage Paper 56. 300 pp.

Allen, W., Richardson, A. (1968). Interaction of light with a plant canopy. J. Opt. Soc. Am. 58:1023-1028.

Allen, W., Gausman, H., Richardson, A. and Thomas, J. (1969). Interaction of isotropic light with a compact plant leaf. J. Opt. Soc. Am. 59: 1376-1379.

Alt, D. Wiemann, F. (1990). Nitrogen in the harvested material and the crop residues of vegetables. Gemuse, 26, 352-356.

ANFFE. 2013/14. Agricultural Fertilizer Consumption. http///www.anffe.com/

AOAC (1990). Official methods of analysis. 15th Ed. Harwitte W. (Ed), pp. 127-129. Association of official analytical chemist. Washington (USA).

Balasubramanian, V., Morales, A.C., Cruz, R.T., Thiyagarajan, T.M., Nagarajan, R., Babu, M., Abdulrachman, S., Hai, L.H., (2000). Adaptation of the chlorophyll meter (SPAD) technology for real-time N management in rice: a review. Int. Rice Res. Inst. 5, 25-26.

Baret, F., Houles, V., Guerif, M. (2007). Quantification of plant stress using remote sensing observations and crop models: the case of nitrogen management. J. Exp. Botany 58, 869-880.

Barker, D.W., Sawyer, J.E. (2010). Using active canopy sensors to quantify corn nitrogen stress and nitrogen application rate. Agron. J., 102, 964-971.

Barnes, E. M., Clarke, T. R., Richards, S. E., Colaizzi, P. D., Haberland, J., Kostrzewski, M.,

& Lascano, R. J. (2000). Coincident detection of crop water stress, nitrogen status and canopy density using ground based multispectral data. In Proceedings of the 5th International Conference on Precision Agriculture, Bloomington, MN (16-19).

Batchelor, W.D., Basso, B., Paz, J.O. (2002). Examples of strategies to analyse spatial and temporal yield variability using crop models. European Journal of Agronomy, 18, 141-158.

Beverly, R.B. (1994). Stem sap testing as a real-time guide to tomato seedling nitrogen and potassium fertilization. Commun. Soil Sci. Plant Anal., 25, 1045-1056.

Bohmer, M., Wiebe, H., Wehrmann, J. (1981). Zur Stickstoffdungung bei Blumenkohl. Gemuse 17, 44-47. Cited in: Pannier, J., Hofman, G., Vanparys, L. (1996). Optimization of a nitrogen advice system: Target values as a function of N- mineralization rates. In Progress in Nitrogen Cycling Studies (pp. 353-358). Springer Netherlands.

Booij, R., Struik, P.C. (1990). Effects of temperature on leaf and curd initiation in relation to juvenility of cauliflower. Scientia Horticulturae,44, 201-214.

Bremner, J.M.   (1997).   Sources   of   nitrous   oxide   in   soils.   Nutrient
        Cyclingin
Agroecosystems, 49, 7-16.

Breschini, S.J., Hartz, T.K. (2002). Pre-sidedress soil nitrate testing reduces nitrogen fertilizer use and nitrate leaching hazard in lettuce production. HortScience 37(7), 1061-1064.

Brisson, N., Gary, C., Justes, E., Roche, D., Zimmer, D., Sierra, J., Bertuzzi, P., Burger, P., Bussiere, F., Cabidoche, Y. M., Cellier, P., Debaeke, P., Gaudillere, J.P., Henault, C., Maraux, F., Seguin, B., Sinoquet, H., (2003). An overview of the crop model STICS. Eur. J. Agron., 18, 309-332.

Campillo, C., Garcia, M. I., Daza, C., Prieto, M. H. (2010). Study of a non-destructive method for estimating the leaf area index in vegetable crops using digital images. HortScience, 45(10), 1459-1463.

Cannavo, P., Recous, S., Parnaudeau, V. Reau, R. (2008). Modeling N dynamics to assess environmental impacts of cropped soils. Advances in Agronomy, 97, 131-174.

Cartelat, A., Cerovic, Z. G., Goulas, Y., Meyer, S., Lelarge, C., Prioul, J.-L., Barbottin, A., Jeuffroy, M.-H., Gate, P., Agati, G., Moya, I. (2005). Optically assessed contents of leaf polyphenolics and chlorophyll as indicators of nitrogen deficiency in wheat (*Triticum aestivum* L.). Field Crops Research, 91, 35-49.

Casa, R., Castaldi, F., Pascucci, S., Pignatti, S. (2014). Chlorophyll estimation in field crops: an assessment of handheld leaf meters and spectral reflectance measurements. J. Agric. Sci.

2014, 1-15.

Cao, Q., Miao, Y., Wang, H., Huang, S., Cheng, S., Khosla, R., Jiang, R. (2013). Nondestructive estimation of rice plant nitrogen status with crop circle multispectral active canopy sensor. Field Crops Research, 151, 133-144.

EEC (1998). Regulation (EC) No. 963/98. Marketing standards for cauliflowers and artichokes. Official Journal of the European Union No. 6 of 8 May 1998.

Cerovic, Z. G., Masdoumier, G., Ben Ghozlen, N., Latouche, G. (2012). A new optical leaf-clip meter for simultaneous non-destructive assessment of leaf chlorophyll and epidermal flavonoids. Physiol. Plant. 2012, 146, 251-260.

Cerovic, Z.G., Ben Ghozlen, N., Milhade, C., Obert, M., Debuisson, S., Le Moigne, M.(2015). Non-destructive diagnostic test for nitrogen nutrition of grapevine (*Vitis vinifera L.*) based on Dualex leaf-clip measurements in the field. Journal of Agricultural and Food Chemistry 63, 3669-3680.

Chishaki, N., Horiguchi, T. (1997). Responses of secondary metabolism in plants to nutrient deficiency. Soil Science and Plant Nutrition, 43, 987-991.

Clevers, J. G. P. W., De Jong, S. M., Epema, G. F., Van Der Meer, F. D., Bakker, W. H., Skidmore, A. K., Scholte, K. H. (2002). Derivation of the red edge index using the MERIS standard band setting. International Journal of Remote Sensing, 23(16), 31693184.

Csizinszky, A.A. (1996). Optimum planting time, plant spacing, and nitrogen and potassium rates to maximize yield of green cauliflower. HortScience, 31, 930-933.

Delgado, J., Follett, R.F., Shaffer, M.J. (2000). Simulation of nitrate-nitrogen dynamics for cropping systems with different rooting depths. Soil Sci. Soc. Am. J., 64, 1050-1054.

DiStefano, J.F., Gholz, H.L. (1986). A proposed use of ion exchange resins to measure nitrogen mineralization and nitrification in intact soil cores. Comm. in Soil Sci. Plant Anal., 17, 989-998.

Doerge, T.A., Roth, R.L., Gardner, B.R. (1991). Nitrogen fertilizer management in Arizona. Univ. of Arizona, College of Agriculture, Rpt. 191025.

Doltra, J, Munoz, P. (2010). Simulation of nitrogen leaching from a fertigated crop rotation in a Mediterranean climate using the EU-Rotate_N and Hydrus-2D models. Agricultural Water Management, 97, 277-285.

Doltra, J., Munoz, P., Anton, A., Arino, J. (2010). Soil and plant nitrogen dynamics of a tomato crop under different fertilization strategies. Acta Hort. (ISHS), 852, 207-214.

Dubrulle, P., Machet, J. M., Damay, N. (2003). Azofert: a new decision support tool for fertiliser N recommendations. 12th Nitrogen Workshop, Exeter, Devon, UK. Wageningen Academis Publishers. pp. 500-501.

El-Shikha, D. M., Waller, P., Hunsaker, D., Clarke, T., Barnes, E. (2007). Groundbased remote sensing for assessing water and nitrogen status of broccoli. Agricultural Water Management, 92, 183-193.

Errebhi, M., Rosen, C.J., Birong, D.E. (1998). Calibration of petiole sap nitrate test for irrigated 'Russet burbank' potato. Commun. Soil Sci. Plant Anal., 29, 23-35.

Estiarte, M., Fililla, I., Serra, J. Penuelas, J. (1994). Effects of nutrient and water stress on leaf phenolic content of peppers and susceptibility to generalist herbivore *Heliocoverpa armigera* (Hubner). Oecologia, 99, 387-391.

Evans, J. R. (1989). Photosynthesis and nitrogen relationships in leaves of C3 plants. Oecologia, 78(1), 9-19.

Everaarts, A.P. (1993). General and quantitative aspects of nitrogen fertilizer use in the cultivation of *Brassica* vegetables. Acta Horticulturae, 339, 149-160.

Everaarts, A.P. (2000). Nitrogen balance during growth of cauliflower. Sci. Hort., 83, 173-186.

Everaarts, A.P., de Moel, C.P. (1991). Growth, development and yield of white cabbage (*Brassica oleracea* var. *capitata*) in relation to time of planting (in Dutch). Verslag 132. PAGV, Lelystad. 50 pp.

Everaarts, A.P., de Moel, C.P. Van Noordwijk, M. (1996). The effect of nitrogen and the method of application on nitrogen uptake of cauliflower and on nitrogen in crop residues and soil at harvest. Netherlands Journal of Agricultural Science, 44, 43-55.

Feller, C., Fink, M. (2002). N-min target values for field vegetables. Acta Horticulturae 571, 195-201.

Fink, M., Scharpf, H.C. (2000). Apparent nitrogen mineralization and recovery of nitrogen supply in field trials with vegetable crops. Journal of Horticultural Science and Biotechnology, 75, 723-726.

Fisk, M.C., Schmidt S.K. (1995). Nitrogen mineralization and microbial biomass nitrogen dynamics in three alpine tundra communities. Soil Sci. Soc. Am. J., 59, 10361043.

Fitzgerald, G.J., Rodriguez, D., Christensen, L.K., Belford, R., Sadras, V.O., Clarke, T.R. (2006). Spectral and thermal sensing for nitrogen and water status in rainfed and irrigated wheat environments. Precision Agric., 7, 233-248.

Galambosova, J., Macak, M., Zivcak, M., Rataj, V., Slamka, P., & Olsovska, K. (2014). Comparison of Spectral Reflectance and Multispectrally Induced Fluorescence to Determine Winter Wheat Nitrogen Deficit. Advanced Materials Research, 1059, 127133.

Garda, M.I., Prieto, M.H., Gonzalez, J.A., Monino, M. J. (2003). Production, quality and nutritional status of a lettuce crop under different production systems in the Guadiana lowlands. Actas de Horticultura (SECH), 39, 374-376.

Gardner, B.R., Roth, R.L. (1989). Midrib nitrate concentration as a means for determining nitrogen needs of broccoli. Journal of Plant Nutrition, 12, 111-125.

Gates, D., Keegan, H., Schleter, J., Weidner, V. (1965). Spectral properties of plants. Appl. Opt. 4: 11-20.

Gates, D. M. (1965). Spectral properties of plants. Applied Optics. 4: 11-20.

Gausman, H. (1974). Leaf reflectance of near-infrared. Photogramm. Eng. 40: 183-191.

Gausman, H., Allen, W. (1973). Optical parameters of leaves of 30 plant species. Plant Physiol. 52: 57-62.

Gausman, H., Allen, W., Cardenas, R. (1969). Reflectance of cotton leaves and their structure. Remote Sens. Environ. 1: 19-22.

Gianquinto, G., Sambo, P., Pimpini, F. (2003). The use of SPAD 502 chlorophyll meter for dynamically optimization the nitrogen supply in potato crop: first results. Acta Horticulturae, 607, 191-196.

Gianquinto, G., Goffart, J.P., Olivier, M., Guarda, G., Colauzzi, M., Dalla Costa, L. Delle Vedove, G., Vos, J., Mackerron, D.K.L. (2004). The use of hand-held chlorophyll meters as tool to assess the nitrogen status and to guide nitrogen fertilization of potato crop. Potato Res., 47, 35-80.

Gianquinto, G., Sambo, P., Borsato, D., (2006). Determination of SPAD threshold values in order to optimise the nitrogen supply in processing tomato. Acta Horticulturae, 700, 159-166.

Giebel, A., Wendroth, O., Reuter, H.I., Kersebaum, K.C., Schwarz, J. (2006). How representatively can we sample soil mineral nitrogen? Journal of Plant Nutrition and Soil Sciences, 169, 52-59.

Gitelson, A., Merzlyak, M. N. (1994). Spectral reflectance changes associated with autumn senescence of *Aesculus hippocastanum* L. and *Acer platanoides* L. leaves. Spectral features and relation to chlorophyll estimation. Journal of Plant Physiology, 143(3), 286-292.

Government of La Rioja (2015). Estad^stica Agraria Regional 2012. Consejena de Agricultura,

Ganadena y Medio Ambiente 2013, 128 pag.

Godoy, L.C.G., Villas Boas, R.L., Bull, L.T. (2003). Utiliza?ao da medida do clorofilometro no manejo da aduba^ao nitrogenada em plantas de pimentao. Revista Brasileira do Ciencia Solo, 27, 1049-1056.

Goffart, J.P., Renard, S., Frankinet, M., Sinnaeve, G., Delvigne, A., Marechal, J. (2006). Leaf chlorophyll content measurements for nitrogen fertilization management of curled-leaves endives in open field. Acta Horticulturae, 700, 207-211.

Goffart, J.P., Olivier, M., Frankinet, M. (2008). Potato crop nitrogen status assessment to improve N fertilization management and efficiency: Past-Present-Future. Potato Research, 51, 355-383.

Goulas, Y., Cerovic, Z.G., Cartelat, A. Moya, I. (2004). Dualex: a new instrument for field measurements of epidermal ultraviolet absorbance by chlorophyll fluorescence. Applied Optics, 43, 4488-4496.

Greenwood, D.J. (2001). Modeling N-response of field vegetable crops grown under diverse conditions with N_ABLE: a review. Journal of Plant Nutrition, 24, 1799-1815.

Greenwood, D.J., Neeteson, J.J. Draycott, A. (1986). Quantitative relationships for the dependence of growth rate of arable crops on their nitrogen content, dry weight and aerial environment. Plant & Soil, 91, 281-301.

Greenwood, D.J., Lemaire, G., Gosse, G., Cruz, P., Draycott, A., Neeteson, J.J.. (1990). Decline in percentage N of C3 and C4 crops with increasing plant mass. Annals of Botany 67: 181-190.

Greenwood, D.J., Rahn, C., Draycott, A., Vaidyanathan, L.V. Paterson, C. (1996). Modelling and measurement of the effects of fertilizer-N and crop residue incorporation on N-dynamics in vegetable cropping. Soil Use and Management, 12, 13-24.

Hatch, D.J., Bhogal, A., Lowell, R.D., Shepherd, M.A. Jarvis, S.C. (2000). Comparison of different methodologies for field measurement of net nitrogen mineralization in pasture soils under different soil conditions. Biol. Fertil. Soils, 32, 287-293.

Hartz, T.K. (1994). A quick test procedure for soil nitrate-nitrogen. Commun. Soil Sci. Plant Anal., 25, 511-515.

Hartz, T.K. (2002a) The assessment of soil and crop nutrient status in the development of efficient fertilizer recommendations. Acta Horticulturae 627, 231-240.

Hartz, T.K. (2002b). Efficient management for cool-season vegetables. Vegetable Research and Information Center. University of California. Davis.

Hartz, T. K. (2003). The assessment of soil and crop nutrient status in the development of efficient fertilizer recommendations. Acta Horticulturae, 627, 231-240.

Hartz, T.K., Lestrange, M., May, D.M. (1993). Nitrogen requirements of drip irrigated peppers. HortScience, 28, 1097-1099.

Hartz, T.K., Bendixen, W.E., Wierdsma, L. (2000). The value of presidedress soil nitrate testing as a nitrogen management tool in irrigated vegetable production. HortScience, 35, 651-656.

Heckman, J.R., Hlubik, W.T., Prostak, D.J., Paterson, J.W. (1995). Pre-sidedress soil nitrate test for sweet corn. HortScience, 30(5), 1033-1036.

Heckman, J.R. (2002). In-season soil nitrate testing as a guide to nitrogen management for annual crops. HortTechnology, 12, 706-710.

Himelrick, D.G., Dozier, Jr. W.A., Wood, C.W., Sharpe, R.R. (1993). Determination of strawberry nitrogen status with SPAD chlorophyll meter. Advances in Strawberry Research, 12, 49-53.

Hochmuth, G.J. (1994). Efficiency ranges for nitrate-nitrogen and potassium for vegetable petiole sap quick tests. HortTechnology, 4, 218-222.

Hochmuth, G.J. (2003). Fertilization of pepper in Florida. Univ. Fla. Fla. Sci. Dept. Circ. 1168. 10 pp. http://edis.ifas.ufl.edu

Hochmuth, G.J. (2009). Plant petiole sap-testing for vegetable crops. Univ. Fla. Fla. Sci. Dept. Circ. 1144. http://edis.ifas.ufl.edu/cv004.

Hochmuth, G.J. (2015). Plant petiole sap testing for vegetable crops. Univ. Fla. Hort. Sci. Dept. IFAS Extension CIR 1144.

Hoel, B.O., Solhaug, K.A. (1998). Effect of irradiance on chlorophyll estimation with the Minolta SPAD-502 leaf chlorophyll meter. Annals of Botany 82, 389-392.

Hoffer, A.M., (1978). Biological and physical considerations in applying computer-aided analysis techniques to remote sensor data, in Remote Sensing: The Quantitative Approach, P.H. Swain and S.M. Davis (Eds), McGraw-Hili Book Company, 227-289.

Hoque, E., Remus, G. (1996). Reflective light properties of tissue layers in Beech (*Fagus sylvatica* L.) leaves. Photochem. Photobiol. 63: 498-506.

Houles, V. (2004). Mise au point d'un outil de modulation intraparcellaire de la fertilisation azotee du ble d'hiver base sur la teledetection et un modele de culture. Saint-Mande (France). Universite de Marne la Vallee. 142 pp.

Huett, D.O., White, E. (1992). Determination of critical nitrogen concentrations of lettuce (*Lactuca sativa* L cv. Montello) grown in sand culture. Australian Journal of Experimental Agriculture, 32, 759-764.

IPCC (2007). Climate Change (2007): The Physical Science Basis. Contribution of Working Group I to the Fourth Assessment Report of the Intergovernmental Panel on Climate Change. Solomon, S., D. Qin, M. Manning, Z. Chen, M. Marquis, K.B. Averyt, M. Tignor H.L. Miller (eds.). Cambridge University Press, Cambridge, United Kingdom and New York, NY, USA, 996 pp. available at: http://www.ipcc.ch/ipccreports/ar4- wg1.htm.

Jeuffroy, M.H., Recous, S. (1999). AZODYN: a simple model simulating the date of nitrogen deficiency for decision support in wheat fertilization, Eur. J. Agron,. 10, 129144.

Jones, J.B., Case, V.W. (1990). Sampling, handling and analyzing plant tissue samples. pp. 389 - 427. In: Westerman, R.L. (Ed.). Soil Testing and Plant Analysis. Third Edition. Soil Science Society of America BookSeries, Number 3. Madison, Wisconsin, USA.

Jones, J.W., Hoogenboom, G., Porter, C.H., Boote, K.J., Batchelor, W.D., Hunt, L.A., Wilkens, P.W., Singh, W., Gijsman, A.J., Ritchie, J.T. (2003). The DSSAT cropping system model. Eur. J. Agron., 18, 235-265.

Jongschaap, R.E.E. (2006). Run-time calibration of simulation models by integrating remote sensing estimates of leaf area index and canopy nitrogen. Eur. J. Agron., 24, 316.

Justes, E., Mary, B., Meynard, J.M., Machet, J.M., Thelier-Huche, L. (1994). Determination of a Critical Nitrogen Dilution Curve for Winter Wheat Crops. Ann. Bot. 74, 397-407.

Justes, E., Jeuffroy, M.H., Mary, B.,   (1997). The nitrogen requirement of major agricultural crops: Wheat, barley and durum wheat. Diagnosis of the nitrogen status in crops. Berlin Heidelberg: Springer-Verlag, (73-89).

Kage, H., Alt, C., Stutzel, H. (2002). Nitrogen concentration of cauliflower organs as determined by organ size, N supply, and radiation environment. Plant and Soil. 246, 201-209.

Keating, B.A., Carberry, P.S., Hammer, G.L., Probert, M.E. (2003). An overview of APSIM, a model designed for farming systems simulation. Eur. J. Agron. 18, 267-288.

Kersebaum, K.C., Hecker, J.M., Mirschel, W., Wegehenkel, M. (eds.) (2007). Modelling water and nutrient dynamics in soil-crop systems. Springer, The Netherlands. 272 pp.

Kiraly, Z. (1964). Effect of nitrogen fertilization on phenol metabolism and stem rust susceptibility of wheat. Journal of Phytopathology, 51, 252-261.

Krusekopf, H.H., Mitchell, J.P., Hartz, T.K., May, D.M., Miyao, E.M., Cahn, M.D. (2002). Pre-sidedress soil nitrate testing identifies processing tomato fields not requiring sidedress fertilizer N. HortScience, 37, 520-524.

Kubota, A., Thompson, T. L., Doerge, T.A., Godin, R.E. (1996). A petiole sap nitrate test for cauliflower. HortScience, 31, 934-937.

Kubota, A., Thompson, T. L., Doerge, T.A., Godin, R.E. (1997). A petiole sap nitrate test for broccoli. Journal of Plant Nutrition, 20, 669-682.

Kumar, L., Schmidt, K., Dury, S. and Skidmore, A. (2001). Imaging spectrometry and vegetation science. Imaging spectrometry: Basic principles and prospective applications. Springer. Dordrecht. pp. 111-155.

Lee, D., Graham, R. (1986). Leaf optical properties of rainforest sun and shade extreme shade plants. Am. J. Bot. 73: 1100-1108.

Lemaire, G., Salette, J. (1984). Relationship between growth dynamics and pre-strike dynamics for a group of grazing grasses. I.-Etude de l'effet du milieu. Agronomie 4: 423-430.

Lemaire, G., Gastal, F., Salette, J. (1989). Analysis of the effect of N nutrition on dry matter yield of a sward by reference to potential yield and optimum N content. Proceedings XVI International Grassland Congress, Nice, France (179-180).

Lemaire, G., F. Gastal (1997). N uptake and distribution in plant canopies. Diagnosis of the nitrogen status in crops. Springer Berlin Heidelberg, 1997. 3-43.

Lemaire, G., Meynard, J. M. (1997). Use of the nitrogen nutrition index for the analysis of agronomical data. In Diagnosis of the nitrogen status in crops (pp. 45-55). Springer Berlin Heidelberg.

Lemaire, G., Jeuffroy, M.H., Gastal, F. (2008). Diagnosis tool for plant and crop N status in vegetative stage. Theory and practices for crop N management. Eur. J. Agron., 28, 614-624.

Lidon, A., Bautista, I., de la Iglesia, F., Oliver, J., Llorca, R., Cruz Romero, G. (2005). Furrow and ridge soil nitrogen mineralization in a surface irrigated artichoke field. Acta Horticulturae, 700, 71-74.

Lisiewska, Z., Kmiecik, W. (1996). Effects of level of nitrogen fertilizer, processing conditions and period of storage of frozen broccoli and cauliflower on vitamin C retention. Food Chemistry, 57, 267-270.

Lopez-Granados, F., Jurado-Exposito, M., Atenciano, S., Garda-Ferrer, A., Sanchez de la

Orden, M., Garda Torres, L. (2002). Spatial variability of agricultural soil parameters in southern Spain. Plant and Soil, 246, 97-105.

Lorenz, H.P., Schlaghecken, J., Engl, G. (1989). Ordnungsgemasse Stickstoffversorgung im Freiland-Gemusebau nach dem "Kulturbegleitenden Nmin- Sollwerte (KNS) - System". Ministerium Landwirtschaft Weinbau Forsten Rheinland. Cited in: Everaarts, A. P. (2000).

MacKerron, D.K.L., Young, M.W., Davies, H.V. (1995). Davies, H.V. (1995). A critical assessment of the value of petiole sap analysis in optimizing the nitrogen nutrition of the potato crop. Plant and Soil, 172, 247-260.

Magdoff, F. (1991). Understanding the Magdoff pre-sidedress nitrate test for corn. Journal of Production Agriculture, 4, 297-305.

Magnifico, V., Lattanzio, V., Sarli, G. (1979). Growth and nutrient removal by broccoli [fertilizer application, uptake]. Journal American Society for Horticultural Science.

MAGRAMA, (2015). Anuario de Estad^stica Agraria. http://www.magrama.gob.es/

Maroto, J.V. (2002). Horticultura herbacea especial. Ed. Mundi-Prensa, 5ª edicion, Madrid. pp. 702.

Martinez, D.E. Guiamet, J.J. (2004). Distortion of the SPAD 502 chlorophyll meter readings by changes in irradiance and leaf water status. Agronomie, 24, 41-46.

McClure, J.W. (1977). The physiology of phenolic compounds in plants. In T. Swain, J.B. Harbourne C.F. Van Sumere, eds. Biochemistry of plant phenolics, Vol. 12. Plenum Press, New York, pp. 525-556.

Meisinger, J.J., Randall, G.W. (1991). Estimating nitrogen budgets for soil-crop systems. In: R.F. Follet, D.R. Keeney and R.M. Cruse (eds.) Managing nitrogen for groundwater quality and farm profitability. SSSA Madison, WI. pp. 85-124.

Miller, R.O. (1998). Extractable nitrate in plant tissue; ion-selective electrode method. p. 85-88. In Kalra, Y.P. (Ed). Handbook of reference methods for plant analysis. 287 pp. Soil and plant analysis council, Inc. Philadelphia.

Mithchell, J., May, D.M., Hartz, T., Miyao, G., Cahn, M., Krusekopf, H.H. (2000). Soil testing to optimize nitrogen management for processing potatoes. Dpt. Of Vegetable   Cropsand WeedScience.   WeedScience   .   UniversityofCalifomia .
        Davis.
http://www.cdfa.ca.gov/is/frep/2000PROC2.doc. 13-01-2003.

Moll, R.H., Kamprath, E.J., Jackson, W.A. (1982). Analysis and interpretation of factors which contribute to efficiency of nitrogen utilization. Agronomy J., 74, 562-564.

Neeteson, J.J. (1995). Nitrogen management for intensively grown arable crops and field vegetables. In: P. Bacon (ed.) Nitrogen Fertilization and the Environment. Pp. 295-325. Marcel Dekker, Inc., New York.

Olasolo, L. (2013). Adjustment, validation and application of the EU-Rotate_N model in an area vulnerable to nitrate contamination. Optimization of nitrogen fertilization (Doctoral Thesis, University of La Rioja).

Olsen, J.K., Lyons, D.J. (1994). Petiole sap nitrate is better than total nitrogen in dried leaf for indicating nitrogen status and yield responsiveness of *capsicum* in subtropical Australia. Aust. J. Experimental Agriculture, 34, 835-843.

Padilla, F. M., Teresa Pena-Fleitas, M., Gallardo, M., Thompson, R. B. (2014). Evaluation of optical sensor measurements of canopy reflectance and of leaf flavonols and chlorophyll contents to assess crop nitrogen status of muskmelon. Eur. J. Agron., 58, 39-52.

Pena, M.T., Thompson, R.B., MarUnez-Gaitan, C., Gallardo, M. Gimenez, C. (2012). Optical systems for monitoring nitrogen status in melon. Proceedings of Horticulture (SECH), 60, 820-824.

Peterson, T.A., Blackmer, T.M., Francis, D.D., Scheppers, J.S., (1993). Using a chlorophyll meter to improve N management. A Webguide in Soil Resource Management: D-13 Fertility. Cooperative Extension, Institute of Agriculture and Natural Resources, University of Nebraska, Lincoln, NE, USA.

Plenet, D. (1995). Functioning of the cultures of doughs under canopy. Determination and application of a nutrition index. Nancy: These de Docteur de l'Institut National Polytechnique de Lorraine.

Prasad, M. Spiers, T.M. (1984). Evaluation of a rapid method for plant sap nitrate analysis. Comm. Soil Sci. Plant Anal., 15, 673-679.

Prasad, M. Spiers, T.M. (1985). A rapid nitrate sap test for outdoor tomatoes. Scientia Horticulturae, 25, 211-215.

Prevot, L., Chauki, H., Troufleau, D., Weiss, M., Baret, F. (2003). Assimilating optical and radar data into the STICS model for wheat crops. Agronomie, 23, 297-303.

Pritchard, K.H., Doerge, T.A., Thompson, T.L. (1995). Evaluation of in-season nitrogen tissue tests for drip irrigated leaf and romaine lettuce. Commun. Soil Sci. Plant Anal., 26, 237-257.

Rahn, C.R., Vaidyanathan, L.V., Paterson, C.D. (1992). Nitrogen residues from *brassica* crops. Aspect App. Bio., 30, 263-270.

Rahn, C.R., Paterson, C.D., Vaidyanathan, L.V.V. (1998). The use of measurements of soil mineral N in understanding the response of crops to fertilizer nitrogen in intensive cropping rotations. J. Agr. Sci., 130, 345-356.

Rahn, C., Zhang, K., Lillywhite, R., Ramos, C., Doltra, J., de Paz, J.M., Riley, H., Fink, M., Nendel, C., Thorup-Kristensen, K., Pedersen, A., Piro, F., Venezia, A., Firth, C., Schmutz, U., Rayns, F., Strohmeyer, K. (2010 a). The development of the EU-Rotate N model and its use to test strategies for N use across Europe. Acta Hort. (ISHS), 852, 73-76.

Rahn, C.R., Zhang, K., Lillywhite, R., Ramos, C., Doltra, J., de Paz, J.M., Riley, H., Fink, M., Nendel, C., Thorup Kristensen, K. Pedersen, A., Piro, F., Venezia, A., Firth, C., Schmutz, U., Rayns, F., Strohmeyer, K. (2010 b). EU-Rotate_N - a European Decision Support System - to Predict Environmental and Economic Consequences of the Management of Nitrogen Fertiliser in Crop Rotations. Europ. J. Hort. Sci., 75, 2032.

Ramos, C. (2005). Soil mineral nitrogen analysis as a guide for nitrogen fertilization of horticultural crops. Proceedings of Horticulture 44, 95-102.

Ramos, C., Agut, A., Lidon, A. L. (2002). Nitrate leaching in important crops of the Valencian Community region (Spain). Environ. Pollut., 118, 215-223.

Ramos, C., Ubeda, S. (2009). Nitrogen fertilization of horticultural crops in the Action Programs for the reduction of nitrate pollution in the different Autonomous Communities. Acts of Horticulture (SECH), 56, 87-92.

Rather, K., Schenk, M.K., Everaarts, A.P., Vethman, S. (1999). Response of yield and quality of cauliflower varieties (*Brassica oleracea* var. *botrytis*) to nitrogen supply. Journal of Horticultural Science and Biotechnology, 74, 658-664.

Rather, K., Manfred, K., Schenk, M.K., Everaarts, A.P., Vethman, S. (2000). Rooting pattern and nitrogen uptake of three cauliflower (*Brassica oleracea* var. *botrytis*) F1- hybrids. Journal of Plant Nutrition and Soil Science, 163, 467-474.

Riley, H., Vagen, I. (2003). Critical N-concentration in broccoli and cauliflower, evaluated in field trials with varying levels and timing of fertilizer. Acta Horticulturae, 627, 241-249.

Rincon, L., Saez, J., Perez, J. A., Gomez, M. D., Pellicer, C. (1999). Growth and nutrient uptake of broccoli. Agr. Agr.: Prod. Prot. Veg,14, 225-236.

Rincon, L., Pellicer, C., Saez, J., Abad^a, A., Perez, A., Mann, C. (2001). Vegetative growth and nutrient uptake of cauliflower. Invest. Agr. Prod. Prot. Veg., 16, 119-130.

Roberts, D.F., Kitchen, N.R., Sudduth, K.A., Drummond, S.T., Scharf, P.C. (2010). Economic and environmental implications of sensor-based nitrogen management. Better Crops, 94, 4-

6.

Rodrigo, M.C. (2006). Rapid nitrate analysis in sap as a tool for the improvement of nitrogen fertilization of artichoke and romanesco. Doctoral thesis. Polytechnic University of Valencia.

Rodrigo, M.C., Ramos, C. (2007a). Chlorophyll measurement as a tool for nitrogen fertilization management in horticultural crops. Actas de Horticultura (SECH), 49, 229-234.

Rodrigo, M.C., Ramos, C. (2007b). Nitrate sap analysis as a tool to assess nitrogen nutrition in artichoke. VI International Symposium on Artichoke, Cardoon and its Wild Relatives. Lorca (Murcia) 28-31 March 2006. Acta Horticulturae, 730, 251-256.

Rodrigo, M.C., Ramos, C. (2007c). Sap Analysis and Leaf Chlorophyll Measurements for Nitrogen Management in Artichoke and Romanesco. 15th Nitrogen Workshop: Towards a better efficiency in N use - Lleida (Spain). A. D. Bosch, M.R. Teira & J.M. Villar (eds.), Editorial Milenio, Lleida (Spain), pp. 57-59.

Rodrigo, M.C., Vano, T., Ramos, C. (2005). Quick sap tests to assess nitrogen status in romanesco (*Brassica oleracea* var *botrytis* L.) plants. In: Proc. 14th N-Workshop on "N management in agrosystems in relation to the Water Framework Directive. 24-26 Oct., Maastricht, J.J. Schroder & J.J. Neeteson (eds). Plant Res. Intntl., Report 116, 272-275.

Ros, G.H., Temminghoff, E.J.M., Hoffland, E. (2011). Nitrogen mineralization: a review and meta-analysis of the predictive value of soil tests. European J. Soil Sci., 62, 162173.

Roujean, J. L., & Breon, F. M. (1995). Estimating PAR absorbed by vegetation from bidirectional reflectance measurements. Remote Sensing of Environment, 51(3), 375384.

Rouse, Jr. J.W., Haas, R.H., Deering, D.W., Schell, J.A., Harlan, J.C. (1973). Monitoring the vernal advancement and retrodegradation (green wave effect) of natural vegetation. Prog. Rep. RSC 1978-1, Remote Sensing Center, Texas AandM Univ., College Station, 93 pp.

Samborski, S. M., Tremblay, N., Fallon, E. (2009). Strategies to make use of plant sensors-based diagnostic information for nitrogen recommendations. Agron. J., 101, 800-816.

Sanchez, C. (1999). Diagnostic tools for efficient nitrogen management of vegetables produced in the low desert. University of Arizona.

Sanchez, C.A. (1998). Diagnostic tools for efficient N management of vegetables produced in the low desert. Ann. Report, California Department of Food and Agriculture Fertilizer Research and Education Program, Sacramento.

Scaife, M.A., Stevens, K.L. (1983). Monitoring sap nitrate in vegetable crops: Comparison of test strips with electrode methods, and effects of time of day and leaf position.

Communications in Soil Science and Plant Analysis 14, 761-771.

Scharf, P.C., Lory, J.A. (2009). Calibrating reflectance measurements to predict optimal sidedress nitrogen rate for corn. Agron. J., 101, 615-625.

Scharpf, H.C. (1991). Stickstoffdungung im Gemusebau, AID-Heft Nr. 1223, Bonn-Bad Godesberg. Cited in Feller and Fink (2002).

Scharpf, H.C., Weier, U. (1996). Investigations on the nitrogen dynamic as a basis for the N fertilizer recommendations in vegetable production. Acta Horticulturae 428, 7383.

Schepers, J.S., Francis, D.D., Vigil, M.F., Below, F.E. (1992). Comparison of corn leaf nitrogen concentration and chlorophyll meter readings. Commun. Soil Sci. Plant. Anal., 23, 2173-2187.

Schroder, J. J., Neeteson, J. J., Oenema, O. Struik, P. C. (2000). Does the crop or the soil indicate how to save nitrogen in maize production. Reviewing the state of the art. Field Crops Research, 66, 151-164.

Sepulveda, J., Garros, V. Ramos, C. (2003). Rapid nitrate analysis in soils and waters. Agricola Vergel, May 2003, 273-278.

Shaffer M.J, Halvorson A.D., Pierce F.J. (1991). Nitrate Leaching and Economic Analysis Package (NLEAP): Model Description and Application. In: Managing Nitrogen for Groundwater Quality and Farm Profitability. R.F. Follett, D.R.Keeney, and R.M. Cruse (eds.). Soil Sci. Soci. of America, Inc. Madison, Wisconsin, USA. 285-322.

Shaffer, M.J., Delgado, J.A., Gross, C.M., Follett, R.F., Gagliard, P. (2010). Simulation Processes for the Nitrogen Loss and Environmental Assessment Package. Advances in Nitrogen Management for Water Quality. Delgado, J.A., R.F. Follett, eds., Ankeny, IA: Soil and Water Conservation Society.

Shaver, T.M., Westfall, D. G., Khosla, R. (2007). Remote sensing of corn N status with active sensors. Western Nutrient Management Conference. 2007. Vol. 7. Salt Lake City, UT. USA.

Skoog, D. A., Holler, F. J., Crouch, S. R. (2007).Instrumental analysis (pp. 477-8). Cengage Learning India.

Smeal, D., Zhang, H., (1994). Chlorophyll meter evaluation for nitrogen management in corn. Commun. Soil Sci. Plant Anal. 25, 1495-1503.

Soil Survey Staff (2006). Keys to Soil Taxonomy. 10th ed. U.S. Department of Agriculture. Natural Resources Conservation Service. 332 pp.

Sripada, R. P., Heiniger, R. W., White, J. G., Meijer, A. D. (2006). Aerial color infrared photography for determining early in-season nitrogen requirements in corn. Agronomy

Journal, 98, 968-977. doi: 10.2134/agronj2005.0200.

Stehfest, E., Bouwman, L. (2006). N2O and NO emission from agricultural fields and soils under natural vegetation: summarizing available measurement data and modeling of global annual emissions. Nutrient Cycling in Agroecosystems, 74, 207-228.

Studstill, D.W., Simonne, E.H., Hutchinson, C.M., Hochmuth, R.C., Dukes, M.D., Davis, W.E. (2003). Petiole sap testing sampling procedures for monitoring pumpkin nutritional status. Comm. Soil Sci. Plant Anal., 34, 2355-2362.

Thompson, R.B., Gallardo, M., Joya, M., Segovia, C., Martinez-Gaitan, C., Granados, M.R. (2009). Evaluation of rapid analysis systems for on-farm nitrate analysis in vegetable cropping. Spanish Journal of Agricultural Research, 7, 200-211.

Tremblay, N (2013). Sensing technologies in horticulture: options and challenges. Chron. Hortic. , 53, 10-14.

Tremblay, N., Wang, Z., Cerovic, Z. G. (2012). Sensing crop nitrogen status with fluorescence indicators. A review. Agron. Sustainable Dev., 32, 451-464.

Tremblay, N., Scharpf, H.C., Weier, U., Laurence, H., Owen, J. (2001). Nitrogen management in field vegetables: A guide to efficient fertilization. Agriculture and AgriFood Canada. 65 pp.

Tremblay, N., Dextraze, L., Roy, G., Belec, C., Charbonneau, F. (2002). Quick nitrogen test for use on bean and sweet corn crops in Quebec (Canada). Acta Horticulturae, 506, 141-146.

Tucker, C.J., (1979). Red and photographic infrared linear combinations for monitoring vegetation. Remote Sens. Environ. 8, 127-150

Ulrich, A. (1952). Physiological bases for assessing the nutritional requirements of plants. Ann. Rev. Plant Physiol., 3, 207-228.

Van Den Boogaard, R., Thorup-Kristensen, K. (1997). Effects of nitrogen fertilization on growth and soil nitrogen depletion in cauliflower. Acta Agriculturae Scandinavica B, Plant Soil Sciences, 47, 149-155.

Van der Burgt, G.J.H.M., Oomen, G.J.M., Habets, A.S.J. Rossing, W.A.H. (2006). The NDICEA model, a tool to improve nitrogen use efficiency in cropping systems. Nutrient Cycling in Agroecosystems, 74, 275-294.

Van Groenigen, J.W., Velthof, G.L., Oenema, O., Van Groenigen, K.J., Van Kessel, C. (2010). Towards an agronomic assessment of N2O emissions: a case study for arable crops. European J. Soil Sci., 61, 903-913.

Vazquez, N., Pardo, A., Suso M.L., Quemada, M. (2006). Drainage and nitrate leaching under processing tomato growth with drip irrigation and plastic mulching. Agriculture, Ecosystems and Environment, 112, 313-323.

Vazquez, N., Pardo, A., Suso, M.L. (2010). Effect of plastic mulch and quantity of N fertilizer on yield and N uptake of cauliflower with drip irrigation. Acta Horticulturae, 852, 325-332.

Villeneuve, S., Coulombe, J., Belec, C., Tremblay, N. (2002). A comparison of sap nitrate test and chlorophyll meter for nitrogen status diagnosis in broccoli (*Brassica oleracea* L. spp *Italica*). Acta Horticulturae, 571, 171-177.

Willstatter, R., Stoll, A. (1918). Untersuchungen uber die Assimilation der Kohlensaure. Verlag von Julius Springer. Berlin. Quoted by Hoque and Remus (1996).

Wehrmann, J., Scharpf, H.C. (1986). The Nmin method an aid to integrating various objectives of nitrogen fertilization. Pflanzenernaehr. Bodenk 149, 428-440.

Wendlandt, W., Hecht, H. (1966). Reflectance spectroscopy. Interscience Publishers. New York.

Westcott, M. P., Wraith, J. M. (1995). Correlation of leaf chlorophyll readings and stem nitrate concentrations in peppermint. Comm. Soil Sci. Plant Anal., 26, 1481-1490.

Westerveld, S. M., McKeown, A. W., Scott-Dupree, C. D., McDonald, M. R. (2003). How well do critical nitrogenconcentrations workforcabbage , carrot, and onion crops?. Hortscience, 38, 1122-1128.

Westerveld, S. M., McKeown, A. W., Scott-Dupree, C. D., McDonald, M. R. (2004). Assessment of chlorophyll and nitrate meters as field-tissue nitrogen tests for cabbage, onions, and carrots. HortTechnology, 14, 179-188.

Wiebe, H.J. (1975). The morphological development of cauliflower and broccoli cultivars depending on temperature. Scientia Horticulturae, 3, 95-101.

Woolley, J. (1971). Reflectance and transmittance of light by leaves. Plant Physiol. 47: 656-662.

Wurr, D.C.E., Akehurst, J.M., Thomas, T.H. (1981). A hypothesis to explain the relationship between low-temperature treatment, gibberellin activity, curd initiation and maturity in cauliflower. Scientia Horticulturae, 15, 321-330.

Youden, W.J. (1950). Index for rating diagnostic tests. Cancer 3, 32-35.

Zhang, K., Greenwood, D.J., Spracklen, W.P., Rahn, C.R., Hammond, J.P., White, P.J., Burns I.G. (2010). A universal agro-hydrological model for water and nitrogen cycles in the soil-crop system SMCR_N: Critical update and further validation. Agric. Water Manage., 97, 1411-

1422.

Zhang, Y., Tremblay, N., Zhu, J. (2012). A First Comparison of Multiplex® for the Assessment of Corn Nitrogen Status. Journal of Food, Agriculture and Environment, 10, 1008-1016.

Annexes

1. Results Barcelona variety, year 2012.

Production (kg/ha)			Pellas	Sheets	Total	
	T1		11.984a	31.903a	43.887a	
	T2		22.142b	46.697b	68.839b	
	T3		23.806b	47.188b	70.994b	
	T4		22.259b	44.954b	67.213b	
			***	***	***	
Total (%)			10-9	1-10	15-10	22-10
	T1		5,11	3,92a	2,29a	2,54a
	T2		5,50	4,65b	3,75b	3,40b
	T3		5,50	4,78b	3,87b	4,02c
	T4		5,66	4,80b	3,75b	4,15c
			ns	***	***	***
N-NO3 sap (ppm)			10-9	1-10	11-10	24-10
	T1		2.615a	690a	13a	6a
	T2		2.145b	2.345b	1.260b	85a
	T3		2,335ab	2.535b	1.660c	695b
	T4		2.275ab	2.515b	2.260d	2.225c
			*	***	***	***
SPAD			10-09	1-10	10-10	23-10
	T1		63,3	58,7	57,4	53,8a
	T2		63,0	56,6	62,6	61,0b
	T3		60,5	60,1	59,3	60,0b
	T4		62,5	59,7	59,7	61,5b
			ns	ns	ns	**
DUALEX (Chl)	T1		10-09	4-10	10-10	23-10
	T2		46,55	43,91a	40,71a	44,74a
	T3		47,64	48,73b	47,21b	51,17b
	T4		46,47	49,63b	47,71b	52,42b
			48,34	51,33b	48,05b	53,53b
			ns	**	***	***
DUALEX (NBI)			10-09	4-10	10-10	23-10
	T1		38,14	27,74a	25,36a	27,15a
	T2		37,80	36,59b	33,59b	36,64b
	T3		39,17	36,97b	34,38b	37,52b
	T4		40,00	40,95b	34,91b	39,02b
			ns	**	***	***

MULTIPLEX (SFR)			10-09	4-10	10-10	23-10
	T1		5.93ab	5,53a	5,15a	5,17a
	T2		5,78a	5,63a	5.36ab	5,54b
	T3		5.91ab	5,74a	5.54bc	5,65b
	T4		6,00b	6,16b	5,69c	5,63b
			**	**	***	***
MULTIPLEX (NBI)			10-09	4-10	10-10	23-10
	T1		1.89ab	1,06a	0,98a	0,86a
	T2		1,81a	1.60bc	1,75b	1,66b
	T3		2,08b	1,48b	1,91b	1,77b
	T4		2.01ab	1,79c	2,01b	1,86b
			**	**	***	***
Crop-Circle (NDRE)				1-10		24-10
	T1			0,329a		0,243a
	T2			0,352b		0,282b
	T3			0,353b		0,312c
	T4			0,350b		0,310c
				***		***
Crop-Circle (NDVI)				1-10		24-10
	T1			0,768a		0,64a
	T2			0,788b		0,68b
	T3			0,789b		0,71b
	T4			0,772b		0,72b
				***		***

Results Barcelona variety, year 2013.

Production (kg/ha)			Pellas	Sheets	Total	
	T1		8.084a	19.105a	27.189a	
	T2		13.316b	27,114ab	40.430b	
	T3		17.116bc	31.422bc	48.538bc	
	T4		20.748c	36.935c	57.683c	
			***	***	***	
Total (%)			26-8	16-9	30-9	21-10
	T1		4,58	3,97a	4,37a	2,64a
	T2		4,34	4.42ab	4.59ab	3,57b
	T3		4,58	4,77b	5,14b	4.20bc
	T4		4,52	5,08b	4,98b	4,72c
			ns	**	**	***
$N\text{-}NO_3^-$ sap (ppm)			28-8	18-9	1-10	21-10
	T1		1.210	483a	228a	2a
	T2		1.231	583ab	798b	16a
	T3		1.287	682b	1.113b	224b
	T4		1.289	659b	1.034b	549c

			ns	***	***	***
SPAD		28-8	4-9	18-9	30-9	21-10
	T1	56,0	61,5	55,1	59,6a	55,3a
	T2	56,4	62,0	55,8	61.0ab	54,8a
	T3	56,6	64,1	54,3	62.6ab	62,0b
	T4	55,6	63,7	55,7	63,8b	64,8b
		ns	ns	ns	*	***
DUALEX (Chl)		28-8	4-9	18-9	30-9	21-10
	T1	40,82	43,39a	39,13	46,68a	40,70a
	T2	40,16	42,37a	38,72	47,88a	44,62a
	T3	41,60	43,77a	38,68	49,98b	48,62b
	T4	40,22	46,16b	41,66	51,67b	53,01c
		ns	***	ns	***	***
DUALEX (NBI)		28-8	4-9	18-9	30-9	21-10
	T1	28,22a	30,41a	26,43a	33,82a	23,62a
	T2	29,12a	30,54a	26,86a	38.68bc	27,07a
	T3	31.04ab	30.64ab	26,47a	42.66bc	31,39b
	T4	29,52b	32,85b	30,08b	44,11c	36,71c
		***	**	**	***	***

MULTIPLEX (SFR)		28-8	4-9	18-9	30-9	21-10
	T1	4.10ab	4,34	4.45ab	4,38a	4,10a
	T2	3,97a	4,31	4,71a	4.53ab	4.23ab
	T3	4.30ab	4,42	4.44ab	4,49a	4,35b
	T4	4,35b	4,37	4,39b	4,65b	4,46b
		*	ns	*	**	**
MULTIPLEX (NBI)		28-8	4-9	18-9	30-9	21-10
	T1	0.98a	1,15a	1,27	1,33a	0,81a
	T2	1.17ab	1,31b	1,45	1,92b	1.02ab
	T3	1,25b	1,29 b	1,41	2,09b	1,31b
	T4	1,31b	1,22a	1,44	2,06b	1,73c
		**	**	ns	***	***
Crop-Circle (NDRE)		30-8		18-9	30-9	
	T1	0,089b		0,181a	0,271a	
	T2	0,102c		0,205b	0,301b	
	T3	0,059[a]		0,214c	0,339c	
	T4	0,109d		0,245d	0,349d	
		***		***	***	
Crop-Circle (NDVI)		30-8		18-9	30-9	
	T1	0,214b		0,447a	0,639a	
	T2	0,255c		0,502b	0,686b	
	T3	0,147a		0,513c	0,758c	

	T4	0,260d		0,582d	0,768d	
		***		***	***	

Results Barcelona variety, year 2014.

Production (kg/ha)			Pellas	Sheets	Total	
	T1		7.328a	24.201a	31.529a	
	T2		11,068ab	30.553b	41.621ab	
	T3		14.517bc	34.952bc	49,470bc	
	T4		17.318c	38.965c	56.283c	
			***	***	***	
Total (%)			10-9	26-9	15-10	3-11
	T1		4,96	3,27a	2,84a	2,27a
	T2		5,18	3,94b	3,91b	2,59a
	T3		5,13	4,32b	4,12bc	3,41b
	T4		4,94	4,11b	4,70c	3,46b
			ns	***	***	***
N-NO3 sap (ppm)			10-9	29-9	17-10	
	T1		1.402	181a	4a	
	T2		1.436	499b	145ab	
	T3		1.375	686b	307b	
	T4		1.465	727b	813c	
			ns	***	***	
SPAD			10-9	29-9	17-10	
	T1		57,4	57,2	54,8a	
	T2		55,5	57,3	57,9b	
	T3		56,3	55,6	59.0bc	
	T4		57,2	55,4	62,5c	
			ns	ns	***	
DUALEX (Chl)			10-9	29-9	17-10	
	T1		45,38	44,09	39,44a	
	T2		44,65	44,01	43,11b	
	T3		45,18	42,11	45,13c	
	T4		44,43	43,44	48,19d	
			ns	ns	***	
DUALEX (NBI)			10-9	29-9	17-10	
	T1		34,73	28,35	24,42a	
	T2		34,84	28,50	28,63b	
	T3		34,79	28,24	32,17c	
	T4		34,90	29,75	36,60d	
			ns	ns	***	
MULTIPLEX (SFR)			10-9	29-9	17-10	
	T1		4,91	4.62ab	4,46a	

	T2			4,93	4,75b	4,71b	
	T3			4,98	4,61a	4,79b	
	T4			5,02	4,92c	4,87b	
				ns	***	***	
MULTIPLEX (NBI)				10-9	29-9	17-10	
	T1			1,47a	0,92a	0,90a	
	T2			1.57ab	1,10b	1,17b	
	T3			1.49ab	1.29bc	1,46c	
	T4			1,65b	1,34c	1,78c	
				*	***	***	
Crop-Circle (NDRE)				10-9	29-9	14-10	
	T1			0,051c	0,198a	0,186a	
	T2			0,045b	0,223b	0,218b	
	T3			0,043a	0,241c	0,251c	
	T4			0,047b	0,243c	0,260d	
				***	***	***	
Crop-Circle (NDVI)				10-9	29-9	14-10	
	T1			0,149c	0,551a	0,593a	
	T2			0,142b	0,585b	0,635b	
	T3			0,138a	0,618c	0,687c	
	T4			0,144b	0,606c	0,673d	
				***	***	***	

Results of Typical variety, year 2013.

Production (kg/ha)				Pellas	Sheets	Total	
	T1			10.035a	32.066a	42.101a	
	T2			14.113ab	37,135ab	51.248a	
	T3			18,910bc	45,635ab	64.545a	
	T4			22.392c	50.811b	73.203b	
				**	**	**	
Total (%)				26-8	16-9	7-10	21-11
	T1			4,47	4,44	3,61a	2,24
	T2			4,65	4,44	4.06ab	2,20
	T3			4,53	4,69	4,56b	2,67
	T4			4,75	4,65	4,60b	3,05
				ns	ns	*	ns
N-NO3 sap (ppm)				29-8	19-9	8-10	4-11
	T1			1.265	413a	57a	2a
	T2			1.350	555b	368a	3a
	T3			1.389	601b	831b	67a
	T4			1.380	840b	962b	230b
				ns	***	***	***
SPAD				29-8	19-9	7-10	6-11

			56,4a	55,9	59,7	58,2a
	T1		56,4a	55,9	59,7	58,2a
	T2		59.3ab	57,3	62,2	62,9b
	T3		58.6ab	55,9	61,8	63,4b
	T4		60,9b	57,7	62,8	64,8b
			*	ns	ns	***
DUALEX(Chl)			29-8	19-9	7-10	6-11
	T1		45,81a	47,17a	44,53a	44,09a
	T2		48,22a	50,02b	47.06ab	45,62a
	T3		48,97a	48,54a	48.31bc	48.61bc
	T4		51,23b	52,20b	49,18c	50,18c
			***	***	***	***
DUALEX (NBI)			29-8	19-9	7-10	6-11
	T1		34,37a	33,62	28,29a	26,28a
	T2		35,37a	35,27	32,54b	28,55a
	T3		35,84a	34,82	36,02c	33,54b
	T4		37,80b	37,37	38,32d	38,40c
			**	ns	***	***

MULTIPLEX (SFR)			29-8	19-9	7-10	6-11
	T1		4,49	4,70a	4,39a	4,04a
	T2		4,77	4.80ab	4.67ab	4,14a
	T3		4,41	4.78ab	4.78bc	4,53a
	T4		4,53	4,87b	4,83c	4,57b
			ns	*	***	***
MULTIPLEX (NBI)			29-8	19-9	7-10	6-11
	T1		1,23	1,20a	0,97a	0,90a
	T2		1,28	1,28a	1,40b	0,93a
	T3		1,25	1.35ab	1,53b	1,74b
	T4		1,23	1,40b	1,78c	1,94b
			ns	*	***	***
Crop-Circle (NDRE)			29-8	19-9	7-10	6-11
	T1		0,031a	0,160b	0,312a	0,246a
	T2		0,030a	0,158b	0,334b	0,273b
	T3		0,031a	0,127a	0,347c	0,320c
	T4		0,035b	0,152b	0,355d	0,339d
			***	***	***	***
Crop-Circle (NDVI)			29-8	19-9	7-10	6-11
	T1		0,099a	0,371b	0,732a	0,696a
	T2		0,100a	0,364b	0,757b	0,721b
	T3		0,094a	0,294a	0,769c	0,759c
	T4		0,104b	0,337b	0,773d	0,766d
			***	***	***	***

Results of Typical variety, year 2014.

Total (%)			1-9	15-9	1-10	23-10
	T1		4,55	2,99a	1,67a	1,97a
	T2		4,25	3,84b	2,57b	2,65a
	T3		4,21	3,92b	3.37bc	3,57b
	T4		4,58	3,97b	3,76c	3,85b
			ns	*	***	***
N-NO3 sap (ppm)			1-9	16-9	1-10	23-10
	T1		1.556	497a	8a	1
	T2		1.522	858b	138a	2
	T3		1.533	1.183c	483b	12
	T4		1.576	1.298c	729b	75
			ns	***	***	ns
SPAD			1-9	16-9	1-10	23-10
	T1		54,5	56,6a	56,1a	57,1a
	T2		55,5	57,7a	55,5a	61.0ab
	T3		56,1	60.9ab	58.2ab	62,8b
	T4		55,7	62,6b	60,5b	64,9b
			ns	**	**	***
DUALEX (Chl)			1-9	16-9	1-10	23-10
			45,48a	45,14a	40,47a	42,36a
			47.22ab	47,92b	42.16ab	48,35b
			47.67bc	49.48bc	44,50b	52,09c
			49,23c	52,17c	48,54c	53,34c
			***	***	***	***
DUALEX (NBI)			1-9	16-9	1-10	23-10
	T1		33,44a	29,41a	22,47a	22,67a
	T2		33,64a	34,32b	26,46b	28,00b
	T3		35.68ab	38,50b	31,94c	32,97c
	T4		37,28b	43,61c	35,32d	34,56c
			***	***	***	***
MULTIPLEX (SFR)			1-9	16-9	1-10	23-10
	T1		5,37	5,10a	4,52a	4,15a
	T2		5,45	5,33b	4,95b	4,46b
	T3		5,30	5.37bc	5,04b	4,73c
	T4		5,13	5,53c	5,43c	5,00d
			ns	***	***	***

MULTIPLEX (NBI)			1-9	16-9	1-10	23-10
	T1		1,48b	1,04a	0,60a	0,41a
	T2		1,39b	1,43b	1,10b	0,94b
	T3		1,49b	1,60b	1,64c	1,51c
	T4		0,94a	1,90c	1,83c	1,48c
			***	***	***	***

Crop-Circle (NDRE)		3-9	16-9	1-10	14-10	23-10
	T1	0,034a	0,154a	0,207a	0,197a	0,156a
	T2	0,049c	0,207b	0,284b	0,262b	0,211b
	T3	0,047c	0,231c	0,342c	0,314c	0,267c
	T4	0,029b	0,209b	0,342c	0,331d	0,273d
		***	***	***	***	***
Crop-Circle (NDVI)		3-9	16-9	1-10	14-10	23-10
	T1	0,118a	0,417a	0,569a	0,608a	0,528a
	T2	0,150b	0,525b	0,710b	0,715b	0,652b
	T3	0,147b	0,571c	0,768c	0,758c	0,722c
	T4	0,116a	0,514b	0,763c	0,764d	0,709c
		***	***	***	***	***

Results of Casper variety, year 2012.

Production (kg/ha)			Pellas	Sheets	Total	
	T1		34.263	64.497	98.760	
	T2		33.287	61.073	94.360	
	T3		34.949	64.683	99.632	
	T4		33.855	62.165	96.019	
			ns	ns	ns	
Total (%)		7-8	22-8	5-9	4-10	8-11
	T1	1,29	5,79	5,49	3,98	3,98
	T2	1,29	5,77	5,48	4,43	4,20
	T3	1,29	5,70	5,85	4,63	3,98
	T4	1,29			4,81	4,18
		ns	ns	ns	ns	ns
N-NO3- sap (ppm)			22-8	4-9	3-10	5-11
	T1		5.363	6.013	1.463	925
	T2		4.850	6.225	2.688	1.413
	T3		4.963	5.588	2.150	1.150
	T4					
			ns	ns	ns	ns

Results of the Casper variety, year 2014.

Production (kg/ha)			Pellas	Sheets	Total	
	T1		26.107a	55.535a	81.642a	
	T2		26.780a	59,660ab	86,440ab	
	T3		30.524b	64,130ab	94.654bc	
	T4		32.253b	66.780b	99.033c	
			**	***	*	
Total (%)			9-9	2-10	20-10	20-10
	T1		5,50	2,85a	2,60	2,23
	T2		5,28	2,78a	2,76	2,46
	T3		5,03	4,02b	3,37	2,82
	T4		5,08	4,30b	3,24	2,99

			9-9	2-10	20-10	
			ns	***	ns	ns
N-NO3- sap (ppm)			9-9	2-10	20-10	
	T1		1.147	13a	4a	
	T2		1.140	167a	16ab	
	T3		1.091	553b	58ab	
	T4		1.154	714b	159b	
			ns	***	*	
SPAD			9-9	2-10	20-10	
	T1		58,05b	51,10	56,64	
	T2		55.40ab	50,94	59,90	
	T3		53,91a	52,71	55,47	
	T4		55.13ab	52,78	59,45	
			*	ns	ns	
DUALEX(Chl)			9-9	2-10	20-10	
	T1		44,07	38,89a	44,82a	
	T2		43,55	40.85ab	46.55ab	
	T3		43,58	41,81b	45,42a	
	T4		44,31	41,22b	47,81b	
			ns	**	**	
DUALEX (NBI)			9-9	2-10	20-10	
	T1		34,25	24,44a	31,10a	
	T2		34,61	25,34a	33.20ab	
	T3		32,95	30,94b	30,11b	
	T4		35,14	33,34b	39.57ab	
			ns	***	***	

			9-9	2-10	20-10	
MULTIPLEX (SFR)			9-9	2-10	20-10	
	T1		1,94	1,66a	1,69a	
	T2		1,92	1,64a	1,69a	
	T3		1,93	1.73ab	1.74ab	
	T4		1,98	1,80b	1,79b	
			ns	**	*	
MULTIPLEX (NBI)			9-9	2-10	20-10	
	T1		0,38	0.30ab	0,37a	
	T2		0,40	0,26a	0,45a	
	T3		0,36	0,44b	0,37a	
	T4		0,41	0,53b	0,60b	
			ns	***	***	
Crop-Circle (NDRE)			9-9	2-10	20-10	
	T1		0,132d	0,266b	0,251a	
	T2		0,108b	0,239a	0,248a	
	T3		0,085a	0,280c	0,270b	
	T4		0,124c	0,297d	0,290c	

Crop-Circle (NDVI)			***	***	***	
			9-9	2-10	20-10	
	T1		0,391d	0,686b	0,663a	
	T2		0,342b	0,641a	0,660a	
	T3		0,304a	0,696c	0,681b	
	T4		0,375c	0,711d	0,701c	
			***	***	***	

Printed by Books on Demand GmbH, Norderstedt / Germany